CHEMISTRY MADE SIMPLE

Revised Edition

CHEMISTRY MADE SIMPLE

By
FRED C. HESS, Ed.D.

Revised by
ARTHUR L. THOMAS, Ph.D.

MADE SIMPLE BOOKS
DOUBLEDAY & COMPANY, INC.
GARDEN CITY, NEW YORK

Library of Congress Cataloging in Publication Data

Hess, Fred C.
  Chemistry made simple.

  Includes index.
  1. Chemistry—Popular works.   I. Thomas, Arthur
Louis.   II. Title.
QD37.H4   1984   540   82-46054
ISBN 0-385-18850-1

ABOUT THIS BOOK

For the person undertaking the study of a branch of science such as chemistry, two skills are essential: to be able to read with comprehension, and to be able to compute mathematical quantities. To ''make chemistry simple,'' an effort has been made to keep the essential skills of the reader in mind, and to offset to the maximum the burden they impose.

The gaining of knowledge through reading is dependent upon one's ability to understand the meaning of both individual words and a sequence or flow of words. To help the reader with the first facet of this problem, words of special chemical meaning have been emphasized in the text and carefully defined, either in a special sentence or by context, when they first appear. As further assistance, a glossary of chemical terms is included in the Appendix. Nevertheless, it is highly recommended that the reader take the time to understand the meaning of new words as they appear.

As to the flow of words, every effort has been made to keep the phraseology and style as simple as possible. Ultimately an author must describe a fact or an idea in a single set of words. Then he can only hope that his message is understandable to the reader.

Arithmetic, it is rumored, is considered by some people to be mean, horrible, and terrifying. Please don't feel that way about it. When used properly, arithmetic can be a powerful tool to help one understand the behavior of nature. Note particularly that arithmetic is not to be an end in itself, but only a means to an end, namely the learning of chemistry.

To reduce the computational burden on the reader, all numerical applications in the text have been worked out in detail. Several problem sets have been included to give the reader an opportunity to try his computational skill, but in each case these problems too have been worked out in detail in the Answer Section. Essentially, all that is required of the reader is that he or she be able to follow the numerical discussion.

The use of an inexpensive hand-held calculator will reduce the computational effort considerably, allowing the reader more time to study the chemical principles illustrated by the numerical problems.

It was felt that to omit arithmetic would be unfair to the reader, for chemistry is a quantitative science. Its omission would have produced a grossly distorted view of elementary chemistry. Therefore the numerical applications have been included so that the reader may use them in the manner he or she chooses.

Read slowly, carefully, thoughtfully. Try the problems. Try the experiments. You will find that elementary chemistry is, indeed, simple.

CONTENTS

LIST OF TABLES

EXPERIMENTS

CHEMISTRY MADE SIMPLE

Revised Edition

CHAPTER 1

INTRODUCTION

Look about you for a moment. What do you see? A painted wall? Is it made of plaster? Or laminated plywood? Or bonded panelboard? Is there a photograph hanging on it? Or a lithographed print? Or an oil painting? Do you see a glass window with nylon curtains and rayon drapes? Is that table by the wall varnished or stained or coated with shellac? Are your chairs covered with plastic? What kind of metal is that lamp made of?

Or look at your clothes. What kinds of dyes have produced those colors? Are the colors ''fast''? And what bleach made those garments white? What does the shop down the street use in cleaning your clothes? Do you use soap in your washing machine? Or detergents? Are your shoes soled with Neolite? Are the heels made of synthetic rubber? Do you have a plastic raincoat? Or has yours been treated to make it water-repellent?

Or look at your kitchen with its porcelain fixtures and waxed floor. Is gas burned in your range? Or does electricity pass through special metals to supply heat for your cooking? Is your kitchen table surfaced with enamel? Or plastic? Is it set with ceramic dishes? Is your cutlery an alloy called sterling silver? Or another alloy called stainless steel? Or is it silver-plated?

As you look about, it rapidly becomes apparent that practically everything you can see is either a product from some chemical industry or has in some way been treated by chemicals. Even the ink and paper of this page are chemical products. Much of your food, your medicines and drugs are either prepared directly in chemical laboratories, or are analyzed by chemists for purity and safety.

As a matter of fact, your body is an astounding chemical factory. It is so complex that chemists have barely started to learn and understand the secrets of the many strange and wonderful processes that take place within it. The body is made up of chemicals which grow! What is perhaps just as remarkable, some of these chemicals grow just so far and then seem to stop. Of course, the growth of chemicals is found in all living things, but the fundamental difference between things that live and things that are inanimate still awaits explanation by scientists.

It is little wonder that people everywhere, as they realize how completely the science of chemistry deals with their daily lives, are becoming more curious as to what this branch of science is all about. As the products of chemistry become more numerous, people are looking for a basis for selecting just the product to suit their needs. They want to know why it is that one type of paint is suitable for their baby's furniture while another may be dangerous. They are disturbed by deposits that form in their steam iron. They read about and discuss civic problems like pollution control. They must select a suitable antifreeze for their cars. They are becoming aware that a knowledge of chemistry will help them deal more effectively with these problems and hundreds like them which come up every day.

SCIENCE AND CHEMISTRY

Science is knowledge gained from the study of the behavior of nature. Chemistry is that branch of science which deals with the composition of all forms of matter and with the change of one form of matter into another. It is concerned not only with **what is** and **what happens** in nature, but also with **how** changes take place. It is this knowledge of how changes occur that leads to our ability to control such changes:

a. By imitating them in making greater quantities of better products,

b. By using them as clues in finding new processes and creating new products,

c. By inhibiting them, at times, to preserve useful products.

THE METHOD OF SCIENCE

As with all branches of science, chemical knowledge is obtained fundamentally by careful observation of the behavior of nature. A particular natural event is called a **phenomenon.** The chemist, in seeking an understanding of how a phenomenon takes place, first uses his training and background in making an intelligent guess as to what really is happening. This guess is called a **hypothesis.** He then undertakes a series of experiments in which he permits the phenomenon to occur over and over again under carefully controlled conditions. On the basis of these tests he gathers evidence which either supports his hypothesis or causes him to reject or modify it. When the weight of evidence seems to indicate that his hypothesis is sound, he then announces his findings as a **theory.** If his theory is subsequently proven to be a non-varying performance in nature, he is said to have discovered a natural **law.** Theories and laws are then applied to other phenomena to increase understanding of them and to adapt them in refining or creating products for man's use.

In the early days of chemistry it was common for one man to carry on this entire process alone. The contributions of these early chemists to man's knowledge and understanding of nature were monumental, and yet, because of the limitations of the conditions under which they worked, erroneous theories and even false laws were generally accepted for a time. As the number of skilled persons carrying on the work of science increased, errors became less common and knowledge of nature was rapidly unfolded. Today it is much more common for teams of skilled chemists to work together on a single project, each member contributing to the solution of the problem from his or her own special background. The work of these teams is called **research.** Basic research is concerned both with the checking and rechecking of apparently sound theories, and with the persistent search for deeper understanding of the behavior of nature. Applied research, on the other hand, is concerned with the application of scientific knowledge to the development of new and better products for consumers.

It is significant that the attitude of scientists is that no theory or law represents absolute truth. They feel, rather, that each new discovery has brought us nearer to truth, but they are always ready to modify their concept of the behavior of nature should conflicting new evidence be uncovered.

The equation for chemical progress might be stated thus:

Careful observation + Persistent search for truth + Intelligent thought = Progress.

This equation represents the scientific method of approaching problems, and has been credited with the tremendous growth of science in the past two centuries.

Just as observation is the starting point that leads to new scientific discoveries, so it can serve as a springboard for undertaking the study of a field of science. It will be suggested frequently during the next chapters that you try certain things. Don't hesitate to try them. They will be simple and harmless, and they may make important contributions to your understanding of chemistry.

MEASUREMENT

One of the basic features of observation is measurement. Modern science simply did not exist until we learned to measure precisely such quantities as distance, volume, weight, temperature, pressure, and time. The invention of suitable devices for measuring these quantities not only enabled scientists to gather quantitative data, but it also permitted the use of mathematical ideas in obtaining real meaning from their observations.

In the case of chemistry, the invention of the balance was a critical development. With it, the most important fact of chemistry yet uncovered could be demonstrated—namely, that all changes in nature from one form to another take place on a definite weight or mass basis. Until this was shown, there simply was no science of chemistry.

THE METRIC SYSTEM

The metric system of weights and measures is used in all scientific work. This is so because:

1. Its parts are fundamentally related. The entire metric system is based upon the length of the meter, a bar of special metal which is carefully preserved at Sèvres, near Paris. This bar was intended to be one ten-millionth of the distance from the equator to the north pole. It is slightly longer than a yard, actually measuring 39.37 inches. It serves as a standard for measuring distance, area, and dry volume.

A cubic centimeter is, for all practical purposes, the same volume as a milliliter, and a thousand milliliters equals one **liter,** which is the standard for liquid measure. A liter is slightly larger than a quart, actually 1.0567 quarts.

One milliliter of water at 39.2° F. has a mass of one **gram,** which is the metric standard for mass. A gram is quite small, about 28 grams equaling one ounce. A nickel coin has a mass of 5 grams.

Mass is proportional to weight. You will often hear "weight" or "weighs" when the actual quantity measured is mass. Mass refers to the amount of matter in a body, whereas weight refers to the gravitational attraction on a body. In this book they are used interchangeably.

Table I
The Metric System

Metric to English	*English to Metric*
LINEAR MEASURE	
1 millimeter (mm.) = 0.03937 inches	1 in. = 25.40 mm.
1 centimeter (cm.) = 0.3937 inches	1 in. = 2.540 cm.
1 meter (m.) = 39.37 inches	1 ft. = 30.48 cm.
1 meter (m.) = 3.281 feet	1 ft. = 0.3048 m.
1 meter (m.) = 1.0936 yards	1 yd. = 0.9144 m.
1 kilometer (km.) = 3281 feet	1 mi. = 1609 m.
1 kilometer (km.) = 0.6214 miles	1 mi. = 1.609 km.
SQUARE MEASURE	
1 sq. cm. = 0.155 sq. in.	1 sq. in. = 6.4516 sq. cm.
1 sq. m. = 10.764 sq. ft.	1 sq. ft. = 0.0929 sq. m.
1 sq. m. = 1.196 sq. yd.	1 sq. yd. = 0.8361 sq. m.
CUBIC MEASURE	
1 cubic centimeter (cc.) = 0.061 cu. in.	1 cu. in. = 16.387 cc.
1 cubic meter (c.m.) = 35.3 cu. ft.	1 cu. ft. = 0.0283 c.m.
1 cubic meter (c.m.) = 1.308 cu. yd.	1 cu. yd. = 0.7645 c.m.
CAPACITY	
1 milliliter (ml.) = 0.0338 fluid oz.	1 fl. oz. = 29.573 ml.
1 liter (l.) = 2.1134 liq. pt.	1 liq. pt. = 0.4732 l.
1 liter (l.) = 1.0567 liq. qt.	1 liq. qt. = 0.9463 l.
1 liter (l.) = 0.9081 dry qt.	1 dry qt. = 1.1012 l.
1 liter (l.) = 0.2642 gal.	1 gal. = 3.7853 l.
WEIGHT	
1 gram (g.) = 15.43 grains (gr.)	1 gr. = 0.0648 g.
1 gram (g.) = 0.0353 ounces	1 oz. = 28.35 g.
1 kilogram (kg.) = 2.2046 pounds	1 lb. = 453.6 g.
1 lb. = 0.4536 kg.	

Table I gives the important conversion factors between the metric system and the English system. In studying this table, try to relate the various quantities to things you are familiar with. For example: Which of your fingernails is 1 centimeter wide? Where is the palm of your hand 10 centimeters wide? Use the Metric-English scale in Figure 1 to measure these and other familiar objects. It may also help you to know that a ¼-teaspoon measuring spoon holds 0.8 milliliters, and that 3 aspirin tablets (5 grains each) have a mass of about 1 gram.

Table II
Metric Prefixes

Used With Basic Units: Meter, Liter, Gram

milli-	0.001
centi-	0.01
deci-	0.1
deka-	10
hecto-	100
kilo-	1000

---

2. **The metric system is very simple.** As with dollars and cents, one may change from one unit to another simply by moving the decimal point. Table II gives the prefixes used in the metric system with each of the basic units.

3. **It is international.** Just as the behavior of nature is independent of national boundaries, so science, the study of natural behavior, is international in scope. Metric units mean the same thing to all people. The English units of our own system are defined by law on the basis of metric units.

A complete description of a standardized metric system recognized worldwide is given by the **International System of Units (SI),** which was adopted in 1960. SI means *Système International.* It is a complete, coherent system of measures. Base SI units include the meter for length, the kilogram for mass, and the second for time.

ON WORKING PROBLEMS

Chemistry is a quantitative science, and much of the real meaning of chemistry is lost if the mathematical relationships are omitted in its study. Actually, the mathematics of chemistry involves only simple arithmetic: addition, subtraction, multiplication, division, simple proportion, raising to powers, and taking roots. We will see that even these last two operations can be simplified. Don't be frightened by numbers. Mathematicians have helped us greatly by giving us numbers that are very easy to work with. Take, for example, a simple problem in multiplication:

$$
\begin{array}{r}
47 \text{ (Multiplicand)} \\
\underline{29} \text{ (Multiplier)} \\
423 \\
\underline{94} \\
1363 \text{ (Product)}
\end{array}
$$

Notice how each number in the multiplier is used with each number in the multiplicand in getting the product. Do you remember Roman numerals? They were once used throughout western

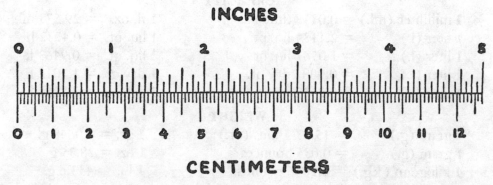

Fig. 1.

civilized countries. Let's look at the same problem in Roman numerals:

$$\begin{array}{r} \text{XLVII} \\ \underline{\text{XXIX}} \\ \text{MCCCLXIII} \end{array}$$

Roman numerals probably retarded the progress of mathematics for centuries.

You will be shown how to go about solving typical chemical problems. Then it will be suggested that you try some practice exercises. Answers to each problem will be supplied. Try them. You will find that they are not too difficult, and they may lead to a much better understanding of chemistry.

SAMPLE PROBLEMS

1. How many inches are there in 15 centimeters?

SOLUTION: In Table I we find that 1 cm. = 0.3937 in. Therefore:

$$15 \; \text{cm.} \times 0.3937 \frac{\text{in.}}{\text{cm.}} = 5.9055 \; \text{in.}$$

Notice that the units must cancel out properly leaving the same units on each side of the equal sign.

2. Express 1273 grams in terms of each of the prefixes in Table II.

SOLUTION:

1,273,000	milligrams	(mg.)
127,300.0	centigrams	(cg.)
12,730.00	decigrams	(dg.)
1,273.000	grams	(g.)
127.3000	dekagrams	(Dg.)
12.73000	hectograms	(Hg.)
1.273000	kilograms	(Kg.)

Notice that in moving in either direction, up or down, the decimal point is changed **one place** for each change in prefix. This is also true if meters or liters replace grams.

3. Express 9255 square meters in terms of each of the prefixes in Table II.

SOLUTION:

9,255,000,000	square millimeters
92,550,000	square centimeters
925,500	square decimeters
9,255	square meters
92.55	square dekameters
0.9255	square hectometers
0.009255	square kilometers

Notice that in working with square measure, the decimal point is moved **two places** for each change in prefix.

4. Express 625 cubic meters in terms of each of the prefixes in Table II.

SOLUTION:

625,000,000,000	cubic millimeters
625,000,000	cubic centimeters
625,000	cubic decimeters
625	cubic meters
0.625	cubic dekameters
0.000625	cubic hectometers
0.000000625	cubic kilometers

Notice that in working with cubic measure, the decimal point is moved **three places** for each change in prefix.

Problem Set No. 1

1. How many milliliters are there in 3 fluid ounces?
2. Convert 1 pound 3 ounces to kilograms.
3. Convert 528 square inches to square meters.
4. Convert 22,400 cubic centimeters to cubic feet.
5. Convert 250 grams to ounces.
6. Convert 25 square yards to square centimeters.
7. How many liters are there in 1 measuring cup (8 fluid oz.)?
8. How many milliliters are there in 1 tablespoon measure (1 tablespoon = 3 teaspoons = 1/16 cup = 1/2 fluid ounce)?
9. A gasoline tank has a capacity of 18 gallons. Express this capacity in liters.
10. Assuming the dimensions of this book to be 8.5 in. × 11 in. × 0.625 in., find the volume of this book in cubic decimeters.

CHAPTER 2

MATTER

Strike a match. Any kind of match will do. Watch it carefully. What do you see? What did you hear? What did you smell? What do you feel? Blow it out. Did it go out completely? Try it again, this time holding the match in a horizontal position. Notice the shape of the flame. Do you see the liquid creeping just ahead of the flame? Light a wooden toothpick with the match, and then blow them both out. Blow harder on the toothpick. Blow it again. What happens? Do you have any evidence that match manufacturers are safety-conscious? What differences in the properties of the match before and after burning can you find? Can the charred remnants of the match still be called a match?

A tremendous amount of chemistry has been illustrated by the phenomena which you have just observed. You will notice that in making observations we use not only our eyes, but also our other senses. The more senses we can employ in observation, the more thorough will be our findings. We will use these observations in becoming acquainted with some of the fundamental terms and ideas of chemistry. The observations will help us visualize chemical ideas and give meaning to the explanation of other phenomena of nature.

PROPERTIES OF MATTER

We distinguish one form of matter from another by its **properties.** When you were asked to handle the match and the toothpick, you knew just what was meant because you were familiar in a general way with the properties of those objects. You are aware, of course, that a wooden match has more properties in common with a toothpick than a paper match. The wood gives the two objects a common substance. A **substance** is a definite variety of matter, all specimens of which have the same properties. Aluminum, iron, rust, salt, and sugar are all examples of substances. Notice that they are all homoge-

neous, or uniform in their makeup. Granite or concrete cannot be called substances because they are not homogeneous. They are made up of several different substances.

Substances have two major classes of properties: physical and chemical. **Physical properties** describe a substance as it is. **Chemical properties** describe the ability of a substance to change into a new and completely different substance.

PHYSICAL PROPERTIES

Substances have two kinds of physical properties: specific and accidental. **Specific physical properties** include those features which definitely distinguish one substance from another. Some of the important specific physical properties are:

Table III
Density of Substances

Substance	g./cc.	lbs./cu.ft.
Aluminum	2.7	168.5
Brass	8.6	536.6
Copper	8.9	555.4
Cork	0.22	13.7
Diamond	3.5	218.4
Gold	19.3	1204.3
Ice	0.917	57.2
Iron	7.9	493.0
Lead	11.3	705.1
Magnesium	1.74	108.6
Mercury	13.6	848.6
Rust	4.5	280.8
Salt	2.18	136.0
Sugar	1.59	99.2
Steel	7.83	488.6
Sulfur	2.0	124.8
Water, fresh	1.0	62.4
Water, sea	1.025	64.0
Zinc	7.1	443.0

1. Density—the mass of a unit volume of a substance. This is usually expressed as g./cc. in the metric system, or lbs./cu.ft. in the English system. Since 1 cc. of water has a mass of 1 g., its density is 1 g./cc. A cubic foot of water has a mass of 62.4 lbs. The density of water in the English system is 62.4 lbs./cu.ft. Multiplying a metric density by 62.4 gives the English density of the substance. Table III lists densities of some common substances.

2. Specific gravity—The ratio of the mass of a given volume of a substance to the mass of the same volume of water at the same temperature. Since 1 cc. of water has a mass of 1 g., specific gravity is numerically equal to the metric density of a substance. Both density and specific gravity have to do with the ''lightness'' or ''heaviness'' of a substance. Aluminum is ''lighter'' than lead. Water is ''lighter'' than mercury. Density is used more with solids, while specific gravity is used more with liquids or solutions (acid in the battery of your car, or alcohol or glycol in the radiator of your car).

3. Hardness—Ability of the substance to resist scratching. A substance will scratch any other substance which is softer. The **MOH hardness scale** is used as a basis for comparing the hardness of substances. This scale is made up of various minerals of different hardness (Table IV), but since so few of these minerals are commonly known, Table IV also gives the approximate hardness of some familiar substances. Low hardness numbers indicate soft substances, and the higher the number, the harder the substance.

4. Odor—Many substances have characteristic odors. Some have pleasant odors, like methyl salicylate (oil of wintergreen); some have pungent odors, like ammonia or sulfur dioxide (a gas which forms when the head of a match burns); some have disagreeable odors, like hydrogen sulfide (a gas which forms in rotten eggs).

5. Color—You are familiar with the color of such substances as gold or copper. White substances are usually described as colorless.

Normally it takes a combination of several specific physical properties to identify a given substance. A single property identifies a substance **only if the property is unique in nature.** Thus, hardness serves to identify the diamond because diamond is the hardest known substance. The color of gold, however, is not unique, as many prospectors unfortunately found out. Their ''strike'' of ''fool's gold'' looked like gold, but turned out to be pyrite, a far less valuable substance also known as iron sulfide.

Accidental physical properties are such features as *weight*, *dimensions*, and *volume*. They have nothing to do with the nature of the substance, but they enable us to find out how much of a given substance we have. **Objects,** particularly manufactured objects, may possess similar accidental properties, but these are in no way fundamentally related to the substances which make up the objects. Thus, matches and toothpicks are objects. Each is made according to a pattern of accidental properties. But toothpicks may be made of wood or of plastic, two completely different substances with totally different specific physical properties.

Table IV
Hardness

MOH Scale		Other Substances	
Talc	1	Graphite	0.7
Gypsum	2	Asphalt	1.3
Calcite	3	Fingernail	1.5
Fluorite	4	Rock salt	2.0
Apatite	5	Aluminum	2.6
Feldspar	6	Copper	2.8
Quartz	7	Brass	3.5
Topaz	8	Knife blade	5.4
Corundum	9	File	6.2
Diamond	10	Glass	6.5

CHEMICAL PROPERTIES

The chemical properties of a substance describe its ability to form new substances under given conditions. A change from one substance to another is called a **chemical change,** or a **chemical reaction.** Hence, the chemical properties of a substance may be considered to be a listing of all the chemical reactions of a sub

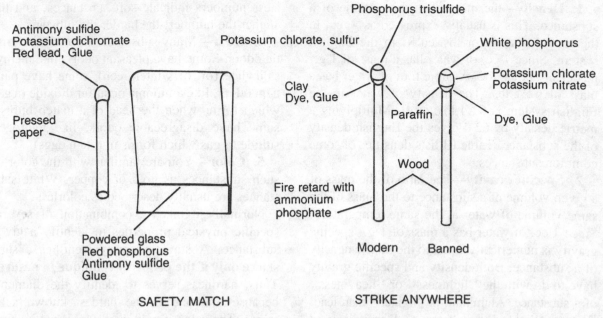

Antimony sulfide
Potassium dichromate
Red lead, Glue

Pressed
paper

Powdered glass
Red phosphorus
Antimony sulfide
Glue

SAFETY MATCH

Phosphorus trisulfide

Potassium chlorate, sulfur

White phosphorus

Clay
Dye, Glue

Potassium chlorate
Potassium nitrate

Paraffin

Dye, Glue

Wood

Fire retard with
ammonium
phosphate

Modern   Banned

STRIKE ANYWHERE

Fig. 2.

stance and the conditions under which the reactions occur.

In the striking of a match, several chemical properties of the substances in a match are illustrated. Examine Figure 2 carefully. Notice the various substances present in each type of match. When you strike a "safety" match, the heat of friction of the head of the match rubbing on the glass is sufficient to cause the phosphorus on the scratching area to burn. This then generates enough heat to cause the substances in the head of the match to ignite. The burning of these, in turn, produces the heat necessary for the matchstick to catch fire. Notice that all of these substances burn (chemical property) but each does so at successively higher temperatures (conditions). None of the substances burns at room temperature! Since the phosphorus is contained only on the scratching area of the box or cover of the matches, they can be "struck" only on this area. (Occasionally safety matches can be struck on glass or linoleum where rubbing produces sufficient heat to cause the head to start burning.)

The phosphorus trisulfide in the tip of the "strike anywhere" match is very sensitive to heat. Rubbing this tip on almost any moderately hard surface will produce sufficient frictional heat to cause this substance to burn. The other substances in the tip, and finally the matchstick, are then ignited as the temperature rises. White phosphorus was formerly used in the tip of this type of match. This substance likewise bursts into flame at temperatures slightly above room temperature. However, the men who worked with white phosphorus and inhaled its fumes contracted a disease known as "phossy jaw" which caused their jaw bones to rot. When laws were passed prohibiting the use of white phosphorus, the company owning the patents on phosphorus trisulfide voluntarily opened them to free public use.

The charred remnants of the matchstick and toothpick consist principally of carbon, one of the new substances formed when wood or paper burn. The "afterglow" you observed in the toothpick is a chemical property of carbon. You have seen the same phenomenon in a charcoal fire. The matchstick exhibited no afterglow because it had been treated with a solution of a fire-retardant substance which soaked into the wood. Borax was formerly used for this purpose, but ammonium phosphate is generally considered to be more effective for this purpose and is now widely used, not only in matchsticks, but also in drapes, tapestries, and other types of decorations.

KINDS OF MATTER

As you look at the different objects about you, you are perhaps impressed by the almost endless variety of matter. Classification of the kinds of matter into fundamental groups was an impossible task until chemists began to probe into the **composition** of matter. Knowledge of composition quickly led to the discovery that all matter is made up of either pure substances or mixtures of pure substances. Substances, in turn, are of two types, either elements or compounds. Figure 3 diagrammatically shows the kinds of matter on the basis of composition.

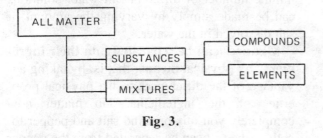

Fig. 3.

ELEMENTS

Elements are the basic constituents of all matter. An element is the simplest form of matter. It cannot be formed from simpler substances, nor can it be decomposed into simpler varieties of matter. Some elements exist free in nature; others are found only in combination. Free or combined, they are the building blocks which make up every different variety of matter in the universe. Table V is a list of the more commonly known elements together with their chemical **symbols.**

How many of these have you seen? How many of them have you heard of? A complete list of the 108 elements known at this time is to be found on page 18.

In general, the symbols are made up of the principal letter or letters in the name of the ele-

Table V

Element	Symbol	Element	Symbol
Aluminum	Al	Neon	Ne
Argon	A	Nickel	Ni
Arsenic	As	Nitrogen	N
Bromine	Br	Oxygen	O
Calcium	Ca	Phosphorus	P
Carbon	C	Platinum	Pt
Chlorine	Cl	Plutonium	Pu
Copper	Cu	Potassium	K
Fluorine	F	Radium	Ra
Gold	Au	Silicon	Si
Helium	He	Silver	Ag
Hydrogen	H	Sodium	Na
Iodine	I	Sulfur	S
Iron	Fe	Tin	Sn
Lead	Pb	Uranium	U
Magnesium	Mg	Zinc	Zn
Mercury	Hg		

ment. The symbols of elements known in antiquity are taken from their Latin names: Copper (Cuprum) Cu; Gold (Aurum) Au; Iron (Ferrum) Fe; Lead (Plumbum) Pb; Mercury (Hydrargyrum) Hg; Potassium (Kalium) K; Silver (Argentum) Ag; Sodium (Natrium) Na; Tin (Stannum) Sn. Symbols are quite important in chemistry, for we will see in the next chapter that they represent more than merely the name of an element.

If all matter were to be broken down into the elements which form it, the percentage of each element in nature would be as shown in Figure 4.

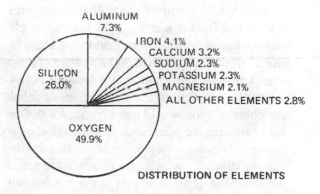

DISTRIBUTION OF ELEMENTS

Fig. 4.

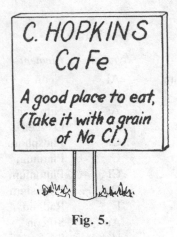

Fig. 5.

The elements in your body can easily be remembered from the advertising sign shown in Figure 5. The symbols of the most common body elements are contained in it. Use Table V to look up the names of the twelve elements represented in the figure. The last two symbols in the sign, NaCl, stand for ordinary table salt.

COMPOUNDS

A compound is a pure substance made up of elements which are chemically combined. They are perfectly homogeneous and have a definite composition regardless of origin, location, size, or shape. A compound can be decomposed into its elements only by some type of chemical change. The elements in a compound cannot be separated by any physical means.

Compounds are much more abundant than elements. About five million compounds are known. Water, sand, ammonia, sugar, salt, alcohol, and benzene are all examples of familiar compounds. It is important to bear in mind that when elements combine to form compounds, the elements lose all of their properties, and a new set of properties unique to the compound is created. For example, if you were to eat any sodium or inhale any chlorine, you would quickly die, for both of these elements are poisonous. But when these two elements combine, they form a compound called sodium chloride, which is ordinary table salt, a substance we must eat as part of our regular diet.

MIXTURES

Most natural forms of matter are mixtures of pure substances. A mixture is a combination of substances held together by physical rather than chemical means. Soil and most rock, plants and animals, coal and oil, air and cooking gas, rivers and oceans, these are all mixtures. Mixtures differ from compounds in the following ways:

The ingredients of a mixture retain their own properties. If you examine a fragment of concrete you will observe that the grains of sand or gravel held together by the cement retain their identity and can be picked free. Their substance has not been changed in the formation of the concrete.

Unlike compounds which have a definite, fixed composition, **mixtures have widely varying composition.** Thus, solutions are mixtures. An infinite number of different salt water solutions can be made simply by varying the amount of salt dissolved in the water.

Mixtures can be separated into their ingredients by physical means, that is, by taking advantage of the differences in the physical properties of the ingredients. No matter how completely you mix or grind salt and pepper together, the salt can be separated from the pepper by dissolving it in water. The insoluble pepper will remain unaffected. The separation is completed by straining or **filtering** the liquid through a piece of cloth which will retain the pepper, and then evaporating the liquid (**filtrate**) to dryness to recrystallize the salt.

Perhaps you would like to try this separation for yourself. Read the following procedure fully and gather your materials before you start. Then proceed with the experiment.

EXPERIMENT 1: Mix a quarter teaspoon of salt and about half that much pepper (ground black) in a small drinking glass. Stir until a good mixture is obtained. Add about a half glass of water and stir until the salt is dissolved. Place a handkerchief or small piece of cloth loosely over the top of a small sauce pan. Filter the liquid into the pan. Notice that all the pepper remains on the cloth and that the salt solution in the pan is perfectly clear. Taste the clear filtrate to see if the salt is really there. Over a very low heat boil away the water in the pan. Be sure to remove the pan just as the last bit of liquid disappears. The white sediment is the recrystallized salt. Taste it to make sure.

The separation of mixtures into ingredients is an important operation. Almost every industry that uses natural products as raw materials employs one or more of the basic methods of separating mixtures. All of the methods take advantage of differences in physical properties of the ingredients. Some of the important methods of separating mixtures are:

Sorting. This involves a selection of the desired ingredient from the waste product in a fragmented mixture. It may be done by hand or by machine. The mining of coal is an example of this process. Here the coal is blasted loose from within the earth and is then separated by sorting from the rock which accompanies the coal.

Magnetic separation. Some iron ore is magnetic. This ore is scooped up in giant shovels from the earth, crushed, and poured onto a magnetized belt as shown in Figure 6. The nonmagnetic waste material drops off the belt at A, but the magnetic ore clings to the belt until it reaches B, and is thus separated.

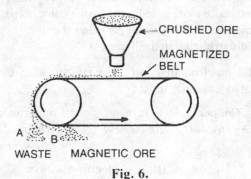

Fig. 6.

Distillation. This process takes advantage of the difference in temperature of boiling **(boiling point)** between the ingredients of a solution. The ingredient with the lowest boiling point boils away first, leaving the higher boiling residue behind. The low boiling ingredient is said to be more **volatile** than the residue. The ingredient which boils off as a gas is then **condensed** back to a liquid by cooling and is collected in a new container.

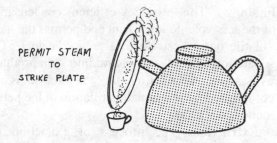

Fig. 7.

EXPERIMENT 2: Dissolve a teaspoon of sugar in a cup of water and place the solution in a tea kettle. Taste the solution to be sure it is sweet. Heat the solution to boiling. Hold a large plate vertically with the far edge just in front of the spout of the kettle so that the steam strikes it. (See Figure 7.) Let the condensed moisture **(condensate)** run down the plate into a cup. Taste the condensate. Is it sweet? Where is the sugar? Which is more volatile, water or sugar? Remove the kettle from the burner before the solution boils completely to dryness.

Simple distillation effectively separates water and sugar because the boiling points of these two substances are relatively far apart. When the boiling points of ingredients to be separated are close together, a process known as **fractional distillation** is used. In this process, a large tower or column is erected above the boiling pot and fitted with cooling coils, or a cooling jacket (See

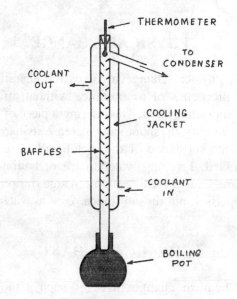

Fig. 8. Fractional Distillation Column

Figure 8). This provides efficient condensation of the less volatile ingredient and permits the more volatile one to escape to a new container. The separation of crude petroleum into such products as gasoline, lubricating oil, and fuel oil is accomplished by fractional distillation of the petroleum.

Extraction. The process of extraction involves the dissolving out of an ingredient from a mixture with a suitable **solvent.** Water was the solvent used to extract salt from the salt and pepper mixture in Experiment 1. Water is also used to extract the flavor of coffee from ground coffee beans in your coffee maker. Alcohol is used to extract vanillin, vanilla flavor, from vanilla beans. Other solvents like benzene, carbon tetrachloride, ether, and acetone are used to extract stains from your clothing.

Gravitation. This process takes advantage of differences in density or specific gravity of the ingredients in a mixture. In the panning of gold, the gold grains settled to the bottom of the pan because of their high density, and the lighter rocks were washed over the edge of the pan with water. In wheat harvesting, the light chaff is blown away from the denser wheat grains. The cleansing action of soap is also based upon this process. Soap bubbles surround the dense dirt particles on your skin or clothing and float the particles away.

PHYSICAL CHANGE

A physical change involves the alteration of the properties of a substance without affecting the substance itself. Hammering a piece of metal will modify its shape and increase its hardness, but the substance of the metal will remain unchanged. Freezing water to ice or boiling it to steam causes a thorough change in physical properties, but the substance remains water.

CHEMICAL CHANGE

Chemical change involves such a thorough change in a substance that an entirely new substance is formed in the process. The new substance created has its own set of properties, so

physical change accompanies chemical change. Do you remember how completely the match was transformed as it burned? That was a chemical change. All burning involves chemical change. So does the rusting of iron, the toasting of bread, the drying of ink, the taking of a photograph, and the digestion of food. Chemical change is a common occurrence.

There are four principal types of chemical change: combination, decomposition, replacement, and double displacement. All chemical changes involve one or a combination of these basic varieties. Let us examine each type more carefully.

1. **Combination.** Combination is the direct joining of two or more simple substances, either elements or simple compounds, to form a more complex compound. For example, copper will join with oxygen in the air when heated to form a compound, copper oxide.

EXPERIMENT 3: Remove about 2 inches of insulation from a 6-inch length of copper wire. Clean the exposed metal with sandpaper to a bright copper color. Heat the copper to redness in the upper part of a gas flame for about one minute. Permit the wire to cool. Notice the black coating on the copper. This is copper oxide. Scrape it off with a knife. This exposes copper metal once more as indicated by the color. Repeat this experiment until you are satisfied that the copper is really **oxidizing** in the flame.

The reaction involved in Experiment 3 can be stated in words thus:

Copper	+	**Oxygen**	=	***Copper oxide**
An element		*An element*		*A compound of the two elements*

2. **Decomposition.** Decomposition is the breaking down of a compound into simpler compounds or into its elements. For example, hydrogen peroxide decomposes in strong light or on

*The equals sign (=) indicates that the substances on the left are transformed into the substance on the right during chemical change. As we will see later, the total weight of substances combining on the left must precisely equal the total weight of products formed on the right. The equals sign emphasizes this quantitative nature of the science of chemistry. Furthermore, many chemical changes are reversible, which means that the substances on the right can be induced to reform the substances on the left. If a reaction goes only one way, from left to right only, then it is irreversible, and instead of writing an equals sign (=), we may write an arrow (→).

contact with skin or other living tissue. Hydrogen peroxide is a compound of hydrogen and oxygen. It decomposes into water, a simpler compound of hydrogen and oxygen, and into oxygen, an element.

EXPERIMENT 4: Pour a small amount of hydrogen peroxide solution into the palm of your hand. Watch the solution closely. The bubbles which form are bubbles of oxygen gas. The rest of the peroxide forms water.

The reaction in Experiment 4 may be stated thus:

Hydrogen peroxide	=	Oxygen	+	Water
A compound		*An element*		*A compound*

3. **Replacement.** Replacement involves the substitution of one element for another in a compound. For example, if a piece of iron were to be dropped into a solution of sulfuric acid (the solution present in the battery of your car), hydrogen gas would be observed bubbling out of the solution. Sulfuric acid is a compound of hydrogen, sulfur, and oxygen. The iron replaces the hydrogen, liberating it as an element, and forms a new compound, iron sulfate (iron, sulfur, and oxygen), in solution. This reaction may be stated thus:

Iron	+	Sulfuric Acid	=
An element		*A compound*	

Hydrogen	+	Iron sulfate
An element		*A compound*

4. **Double replacement.** In double replacement reactions, two compounds react to form two new compounds by exchanging parts. To observe a reaction of this type we need a special solution. Let us make it first.

EXPERIMENT 5: Phenolphthalein is the active ingredient of many common laxatives. It can be extracted from them as follows. Crack and peel off the sugar coating from two Feen-a-mint* tablets, taking care to disturb as little as possible the yellow powder just under the coating. Place the two tablets in a small cup and add one tablespoon of rubbing alcohol. Stir until the yellow phenolphthalein is dissolved from the gum, forming a pale yellow solution.

Keep this phenolphthalein solution in a stop-

*Trade name.

pered bottle. An old well-rinsed nose-drop bottle would be excellent. We will use this solution several times. Phenolphthalein has the property of turning red in solutions of alkalis, but is colorless in acid solutions. An **alkali** is a compound which is the opposite of an **acid.** An alkali **neutralizes** an acid to water and salt solution. Such a reaction is a double displacement type. Let us observe one.

EXPERIMENT 6: Dissolve a few crystals of caustic soda (sodium hydroxide) in one quarter cup of water. Add 2 or 3 drops of phenolphthalein solution prepared in Experiment 5. The red color shows that sodium hydroxide is an alkali. Add vinegar (acetic acid) drop by drop with stirring to the sodium hydroxide solution. When the phenolphthalein becomes colorless, the reaction is completed.

All of the substances involved in the reaction in Experiment 6 are compounds. The reaction may be stated thus:

Sodium Hydroxide	+	Acetic Acid	=	Sodium Acetate	+	Water
An alkali		*An acid*		*A salt*		

In every chemical change, energy is either given off or absorbed. **Energy** is the ability to do work. Heat, light, sound, and electricity are some of the many forms of energy. Fuel oil burns to produce heat. Magnesium burns in a flashbulb to produce light. Dynamite explodes to produce sound and shock. On the other hand, water decomposes into its elements, hydrogen and oxygen, by absorbing electrical energy. A photograph is made by the absorption of light by the chemicals in the film.

It is important not to confuse the energy change in a reaction with the conditions under which a reaction occurs. Wood burns to produce heat, but not at room temperature. The wood must first be heated to a point considerably above room temperature before it will begin to burn. The high initial temperature is a condition under which the reaction of burning takes place. The production of heat by burning wood is a result of the reaction itself.

Many reactions take place only in the presence of a **catalyst.** A catalyst is a substance which alters the speed or rate of a chemical without becoming permanently changed itself. A catalyst which slows down a reaction is called a **negative catalyst** or **inhibitor.** Water is a catalyst for many reactions. Perfectly dry iron will not rust in dry air. Dry crystals of acetic acid will not react as in Experiment 6 with dry crystals of sodium hydroxide.

EXPERIMENT 7: With a match, try to burn a cube of sugar. Notice that the sugar melts but does not burn. Dip the other end of the sugar cube into some cigarette or cigar ashes. *Bear in mind that these ashes have already been burned!* Apply a flame to the ash-covered end of the cube. It now burns because of the presence of a catalyst.

SAMPLE PROBLEMS

1. How much would 400 cc. of mercury weigh?

SOLUTION: From Table III: density of mercury is 13.6 g./cc.

Therefore:

$$400 \text{ cc.} \times 13.6 \text{ g./cc.} = 5440 \text{ g.}$$

2. A solution has a specific gravity of 1.20. How many cc. would 10 g. of this solution occupy?

SOLUTION: A specific gravity of 1.20 = a density of 1.20 g./cc.

Therefore:

$$\frac{10 \text{ g.}}{1.20 \text{ g./cc.}} = 8.33 \text{ cc.}$$

Problem Set No. 2

1. Which of the following physical properties of water are specific and which are accidental?
    (a) Water freezes at 32° F.
    (b) A sample of spring water has a temperature of 46° F.
    (c) Water dissolves in alcohol.
2. A sample of aluminum has a mass of 5.4 g. What is its volume?
3. How many cc. of cork would have the same mass as 1 cc. of gold?
4. The acid in a car battery has a specific gravity of 1.28 when the battery is fully charged. How much would 1 cubic foot of this acid solution weigh?
5. (a) Which substances in Table III would float on water?

(b) Which substances in Table III would sink in mercury?
6. Name the body elements listed in Figure 5.
7. Which of the following are mixtures and which are substances?
    (a) Ocean water.
    (b) A snowflake.
    (c) Ink.
    (d) Paint.
    (e) Sugar.
8. Which of the following are physical changes and which are chemical changes?
    (a) Snow melting.
    (b) Milk souring.
    (c) Sodium bicarbonate neutralizing an "acid" stomach.
    (d) Gas bubbling out of ginger ale.
    (e) The disappearance of sugar when it dissolves in water.
9. Which type of chemical change is illustrated by the following?
    (a) The rusting of iron.
    (b) The explosion of dynamite.
    (c) Acid "eating" a piece of zinc.
    (d) Boric acid or vinegar solutions "treating" a lye burn on the skin.
10. What was the catalyst which aided the decomposition of hydrogen peroxide in Experiment 4? (Where did the bubbles of oxygen form first?)

SUMMARY

Matter is anything that has mass and occupies space. It has both **chemical** and **physical properties.**

Physical properties describe matter as it is.

Chemical properties describe the ability of a form of matter to change to new substances.

Matter is made up of **elements, compounds,** and **mixtures.**

Physical change involves modification of properties without changing the substance.

Chemical change involves forming new substances.

Energy changes accompany chemical changes.

Catalysts change the speed of a chemical reaction.

CHAPTER 3

STRUCTURE OF MATTER

We have seen that chemical change involves a complete transformation of one substance into another. Early chemists reasoned that such a thorough change must in some way be related to the way matter is constructed. They sought to find out the nature of the building blocks which made up the different varieties of matter. They hoped that once they could create some sort of "model" of the fundamental particles of matter, they could then explain not only the various ways that matter was constructed, but also the behavior of substances during the process of chemical change.

As early as 450 B.C. the Greek philosophers reasoned that all matter was built up of tiny particles called **atoms.** Development of this idea was slow, but in 1802 Dalton suggested that all matter could be broken down into elements, the smallest particles of which he referred to as atoms. By 1895, the theory that atoms existed was extended to account for particles of matter even smaller than atoms. By 1913, evidence of the presence of several subatomic particles had been gathered. The work of probing into the structure of matter continues at the present moment. We have not yet learned the full story, and many features of the behavior of matter are still unexplained. There is much room in the field of science for young people with talent.

From a chemical point of view an atomic model has been developed which is quite satisfactory. We will use it to explain all common phenomena. We will also look at some of its weaker points in order to show that science is not cut and dried, but rather is constantly changing as scientists progress toward a better understanding of nature.

ATOMS

If we were to take a strip of aluminum (an element) or a piece of copper (an element) and subdivide it into smaller and smaller pieces, we would eventually come to a tiny particle which, if further subdivided, would no longer show the properties of the element. We call the smallest particle of an element which has all the properties of the element an **atom.** Atoms are really quite small but microscopes are being developed to see them. It would take about 100 million atoms to make a line one inch long. You can thus see that a one inch cube would contain a fantastic number of atoms. The important thing is that atoms are both small and numerous.

ATOMIC STRUCTURE

In 1913 Neils Bohr, a Danish scientist, suggested an atomic model which serves chemists well to the present day. He pictured the atom as consisting of three basic kinds of particles: **electrons, protons,** and **neutrons.** The electron is a particle possessing a negative (−) electrical charge. The proton is a particle consisting of a positive (+) electrical charge equal in magnitude (but opposite in type) to the charge on the electron. The neutron is a particle with no electrical charge. The proton and neutron have essentially the same weight. A weight of one unit has been assigned to each. The electron is much smaller, weighing about 1/1848 times as much as either of the other two. From a chemical point of view, we can consider the weight of the electron to be zero. Table VI summarizes the properties of these three particles which make up an atom.

Table VI		
Particle	*Charge*	*Weight*
Electron	−1	0
Proton	+1	1
Neutron	0	1

In the Bohr model of the atom, protons and neutrons are considered to be packed together in the center of the atom to form what is known as the **nucleus.** Electrons travel about this nucleus in orbits which are at relatively large distances from the nucleus. The average nucleus occupies about one-ten thousandth of the total volume of an atom. The situation is quite similar to the planets revolving about the sun in our solar system.

At this point, three important characteristics of atoms can be stated:

1. Despite the presence of electrically charged particles in atoms, all elements are observed to be electrically neutral. Therefore, the number of positive protons in the nucleus of an atom must be equal to the number of electrons surrounding the nucleus.

2. Since elements differ from one another, their atoms must differ structurally. Each element has an **atomic number.** The atomic number is more than just a catalog number. It is a special characteristic of each element. In the Bohr model, the atomic number is equal to the number of electrons revolving about the nucleus of the atom. Thus, each atom of hydrogen (atomic number 1) has a single electron spinning about the hydrogen nucleus. Each atom of uranium (atomic number 92) has 92 electrons spinning about the uranium nucleus. Since atoms are electrically neutral, the atomic number also equals the number of protons present in the nucleus of an atom.

3. Equal numbers of atoms of different elements weighed under the same conditions have a different weight. Therefore, the atoms of different elements have different atomic weights. The **atomic weight** of an **atom** is equal to the sum of the number of protons and the number of neutrons in the nucleus of the atom. Thus, all of the weight of an atom comes from its nucleus. Atomic weights are relative, which is to say they do not give the number of grams or pounds that an atom weighs, but they merely tell how much heavier or lighter an atom of one element is than another. For example, the atomic weight of oxygen is 16 and the atomic weight of helium is 4. This means that each atom of oxygen weighs 16/4, or 4 times as much as each atom of helium.

These three atomic characteristics are summarized in Table VII.

It may be well to pause here to see how our atomic model is shaping up. Can you visualize a nugget or kernel like a popcorn ball with tiny specks of dust spinning round and round it? Perhaps the popcorn ball also has peanuts in it, giving it two different kinds of particles. We can think of the popcorn as protons and the peanuts as neutrons, all tightly held together in the nucleus. The specks of dust spinning around would be the electrons, equal in number to the pieces of popcorn. The specks of dust would contribute practically nothing to the total weight of our imaginary atom. A model of hydrogen would consist of a single piece of popcorn with a single speck of dust spinning around it. A uranium model would contain quite a lot of popcorn (92 pieces) and many peanuts (146). It would also be quite dusty (92 specks). The sum of the particles in the uranium nucleus is 238, which is the atomic weight of uranium. Figure 9 gives us another picture of our model.

Table VII

Characteristic	Structural Explanation
Neutral atoms	Number of electrons = Number of protons
Atomic Number	Number of electrons = Number of protons = atomic number
Atomic Weight	Number of protons + Number of neutrons = atomic weight

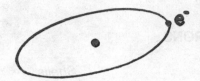

HYDROGEN

At. No. 1

At. Wt. 1

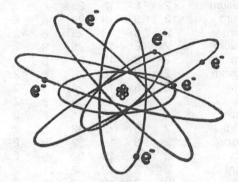

CARBON

At. No. 6

At. Wt. 12

Fig. 9.

DISTRIBUTION OF ELECTRONS

Our model of an atom is still incomplete. The electrons which revolve about the nucleus do so according to a definite pattern. Groups of electrons maintain definite average distances from the nucleus, thereby forming what may be called **shells** of electrons surrounding the nucleus. Each shell is capable of containing a definite number of electrons, the number increasing as the distance from the nucleus increases. The shells are designated by letters—k, l, m, n, o, p—starting with the shell nearest the nucleus. The k-shell can contain up to 2 electrons, the l-shell up to 8, the m-shell up to 18, and the n-shell up to 32. The maximum number of electrons in any shell can be calculated from the relationship:

$$Number = 2 s^2 \qquad (1)$$

where:

Number = maximum number of electrons possible in the shell.

s = the number of the shell (k = 1, l = 2, m = 3, etc.).

The distribution of electrons by shells for the atoms of each element is given in Table VIII. As you read through this list starting with element number 1, hydrogen, be sure to notice the following points:

1. In the first 18 elements the new electron is always added in the outermost shell until the shell is filled. Then a new shell is started.

2. In the higher numbered elements there can be 2 or even 3 unfilled shells of electrons.

3. Eight electrons temporarily fill each of the shells beyond the m-shell, and a new shell must be started before more electrons can be fitted into the temporarily filled shell.

4. **There are never more than 8 electrons in the outermost shell.**

On the basis of the distribution of electrons we can detect four different structural types of atoms in Table VIII. These are:

1. **Noble gas elements**—Those with all shells filled. (These are underlined in Table VIII.) They are sometimes included with the simple elements, as shown in Figure 12.

2. **Simple elements**—Those with only one unfilled shell.

3. **Transition elements**—Those with two unfilled shells.

4. **Inner transition elements**—Those with three unfilled shells. These are the lanthanide and actinide elements.

At first glance, this whole problem of the distribution of electrons in our atomic model might appear to be quite imposing. Actually it is not as hard as it may seem. Remember that we want a model that is useful in explaining chemical change. Two vitally important points are basic in relating chemical change with atomic structure. These are:

Only electrons are involved in chemical change. The nuclei of atoms are in no way altered during chemical change.

In particular, **the electrons in the outermost shell are affected during chemical change.** Occasionally electrons from the second outermost

TABLE VIII
DISTRIBUTION OF ELECTRONS

At. No.	Element	Shells				
		k	l	m	n	o
1	Hydrogen	1				
2	Helium	2				
3	Lithium	2	1			
4	Beryllium	2	2			
5	Boron	2	3			
6	Carbon	2	4			
7	Nitrogen	2	5			
8	Oxygen	2	6			
9	Fluorine	2	7			
10	Neon	2	8			
11	Sodium	2	8	1		
12	Magnesium	2	8	2		
13	Aluminum	2	8	3		
14	Silicon	2	8	4		
15	Phosphorus	2	8	5		
16	Sulfur	2	8	6		
17	Chlorine	2	8	7		
18	Argon	2	8	8		
19	Potassium	2	8	8	1	
20	Calcium	2	8	8	2	
21	Scandium	2	8	9	2	
22	Titanium	2	8	10	2	
23	Vanadium	2	8	11	2	
24	Chromium	2	8	13	1	
25	Manganese	2	8	13	2	
26	Iron	2	8	14	2	
27	Cobalt	2	8	15	2	
28	Nickel	2	8	16	2	
29	Copper	2	8	18	1	
30	Zinc	2	8	18	2	
31	Gallium	2	8	18	3	
32	Germanium	2	8	18	4	
33	Arsenic	2	8	18	5	
34	Selenium	2	8	18	6	
35	Bromine	2	8	18	7	
36	Krypton	2	8	18	8	
37	Rubidium	2	8	18	8	1
38	Strontium	2	8	18	8	2
39	Yttrium	2	8	18	9	2
40	Zirconium	2	8	18	10	2
41	Niobium	2	8	18	12	1
42	Molybdenum	2	8	18	13	1
43	Technetium	2	8	18	13	2
44	Ruthenium	2	8	18	15	1
45	Rhodium	2	8	18	16	1
46	Palladium	2	8	18	18	0
47	Silver	2	8	18	18	1
48	Cadmium	2	8	18	18	2
49	Indium	2	8	18	18	3
50	Tin	2	8	18	18	4
51	Antimony	2	8	18	18	5
52	Tellurium	2	8	18	18	6
53	Iodine	2	8	18	18	7
54	Xenon	2	8	18	18	8

At. No.	Element	Shells						
		k	l	m	n	o	p	q
55	Cesium	2	8	18	18	8	1	
56	Barium	2	8	18	18	8	2	
57	Lanthanum	2	8	18	18	9	2	
58	Cerium	2	8	18	20	8	2	
59	Pra'mium	2	8	18	21	8	2	
60	Neodymium	2	8	18	22	8	2	
61	Promethium	2	8	18	23	8	2	
62	Samarium	2	8	18	24	8	2	
63	Europium	2	8	18	25	8	2	
64	Gadolinium	2	8	18	25	9	2	
65	Terbium	2	8	18	27	8	2	
66	Dysprosium	2	8	18	28	8	2	
67	Holmium	2	8	18	29	8	2	
68	Erbium	2	8	18	30	8	2	
69	Thulium	2	8	18	31	8	2	
70	Ytterbium	2	8	18	32	8	2	
71	Lutetium	2	8	18	32	9	2	
72	Hafnium	2	8	18	32	10	2	
73	Tantalum	2	8	18	32	11	2	
74	Tungsten	2	8	18	32	12	2	
75	Rhenium	2	8	18	32	13	2	
76	Osmium	2	8	18	32	14	2	
77	Iridium	2	8	18	32	15	2	
78	Platinum	2	8	18	32	16	2	
79	Gold	2	8	18	32	18	1	
80	Mercury	2	8	18	32	18	2	
81	Thallium	2	8	18	32	18	3	
82	Lead	2	8	18	32	18	4	
83	Bismuth	2	8	18	32	18	5	
84	Polonium	2	8	18	32	18	6	
85	Astatine	2	8	18	32	18	7	
86	Radon	2	8	18	32	18	8	
87	Francium	2	8	18	32	18	8	1
88	Radium	2	8	18	32	18	8	2
89	Actinium	2	8	18	32	18	9	2
90	Thorium	2	8	18	32	18	10	2
91	Pr'tinium	2	8	18	32	20	9	2
92	Uranium	2	8	18	32	21	9	2
93	Neptunium	2	8	18	32	22	9	2
94	Plutonium	2	8	18	32	24	8	2
95	Americium	2	8	18	32	25	8	2
96	Curium	2	8	18	32	25	9	2
97	Berkelium	2	8	18	32	27	8	2
98	Californium	2	8	18	32	28	8	2
99	Einsteinium	2	8	18	32	29	8	2
100	Fermium	2	8	18	32	30	8	2
101	Mendelevium	2	8	18	32	31	8	2
102	Nobelium	2	8	18	32	32	8	2
103	Lawrencium	2	8	18	32	32	9	2
104	Rutherfordium	2	8	18	32	32	10	2
105	Hahnium	2	8	18	32	32	11	2
106	(Not yet named)	2	8	18	32	32	12	2
107	(Not yet named)	2	8	18	32	32	13	2
109	(Not yet named)	2	8	18	32	32	15	2

shell may be affected in some of the higher numbered elements, but the influence of chemical change never penetrates the atom deeper than the second outermost shell.

ISOTOPES

Evidence is available to show that not all of the atoms of a given element are identical. They may vary in atomic weight. Atoms of an element with different atomic weights are called **isotopes** of the element. About 2,000 isotopes are known, of an estimated 8,000 it is believed exist.

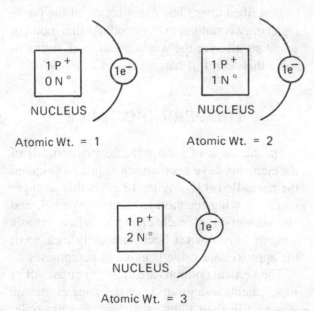

Fig. 10. The 3 Isotopes of Hydrogen

Examine Figure 10. This shows three different kinds of hydrogen atoms. The first has an atomic weight of 1, the second has an atomic weight of 2, and the third has an atomic weight of 3. Notice that the only structural difference is the number of neutrons in the nucleus of each isotope. All three isotopes have but one electron because all are atoms of hydrogen and have atomic number 1. Similarly, all three isotopes have a single proton in the nucleus because each must remain electrically neutral. Isotopes, then, are atoms of the same element possessing different numbers of neutrons in their nuclei.

ATOMIC WEIGHTS

Almost all of the elements have isotopes. The relative abundance of each isotope of a given element in nature varies considerably. For example, the element chlorine has two principal isotopes, one of atomic weight 35 and one of atomic weight 37. If you were to pick up a container of chlorine can be found in the table on page 21. You tainer would have atomic weight 35, and the other 25% would have atomic weight 37. The average mass of all the atoms in the container would then be about 35.5. The listed atomic weight of chlorine can be found in the table on page 21. You will find it to be 35.45. This number is the average atomic weight of all the atoms present in a sample of natural chlorine. The listed **atomic weight** of any **element** is the average of the atomic weights of the isotopes of the element, taking into account the relative abundance of each isotope in a natural sample.

On a practical basis, the average atomic weight of an element is measured by comparing the mass of a given number of atoms of the element to the mass of the same number of atoms of oxygen. The mass of carbon is taken as 12. How chemists know when they are dealing with a given number of atoms will be described later.

SYMBOLS

In Table V (Chapter 2) the symbols of some of the more common elements were given. These symbols are very important in chemistry for they represent three things:
1. The name of an element.
2. One atom of an element.
3. A quantity of the element equal in weight to its atomic weight.

For example, when we write the symbol O, we mean not only the name, oxygen, but we also represent a single oxygen atom with this symbol. What is perhaps most important of all, since oxygen has an atomic weight of 16, the symbol O stands for 16 units of weight of this element. This may be 16 grams, or 16 pounds, or 16 tons. We can select any system of weight units we need when we use symbols to indicate quantities of elements. This idea will be developed further in the next chapter.

THE BOHR MODEL

Our atomic model, as created by Bohr, is now sufficiently developed to explain chemical phenomena. It contains a nucleus composed of positive protons and neutral neutrons which supply the weight of the atom. Surrounding the nucleus are shells of electrons carrying sufficient negative charge to offset the positive charge on the nucleus. Figure 11 is a diagram of the atomic structure of an isotope of phosphorus of atomic weight 31. It shows the number of protons and neutrons in the nucleus, and the number and distribution of electrons in the shells. The atomic weight is the sum of the number of protons and neutrons in the nucleus. The atomic number is the sum of the electrons in the shells.

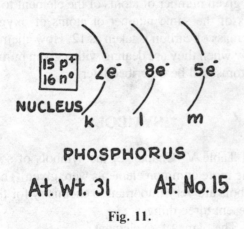

PHOSPHORUS

At. Wt. 31   At. No. 15

Fig. 11.

If you are familiar with the properties of electricity, you know that opposite charges attract one another and like charges repel one another. As you look at Figure 11, two questions might be raised.

Why aren't the negative electrons attracted into the positive nucleus, causing our model to collapse? The answer to this is in the idea that the electrons are spinning about the nucleus. If you tie a piece of string to a ball and whirl it around, the string will get tight. The whirling motion of

the ball causes it to want to fly away from your hand, but the string holds it back. In our atom, the electrical attraction of the positive nucleus and the negative electron just balances the tendency of the whirling electron to escape from the nucleus.

A second question suggested by Figure 11 is: Why doesn't the nucleus fly apart as a result of the repulsion of the protons on one another? In answer to this we can merely state that there is some sort of packing energy holding nuclei together. This energy is not always 100% efficient, because we know that some nuclei do break apart in a process known as **radioactivity.** This will be described later. The exact nature of the packing energy is not yet understood. At this moment scientists all over the world are at work trying to solve this secret of nature.

THE PERIODIC TABLE

On the basis of electronic distribution, all of the elements have been arranged in a table called the **periodic table.** Figure 12 gives this arrangement, showing the atomic number, symbol, and atomic weight of each element. Where atomic weights have not yet been accurately measured, the approximate value is given in parentheses.

The vertical columns are called **groups.** All of the elements in a group have the same electronic structure in their outermost shell. For example, all of the elements in Group I have 1 electron in the outermost shell. (Check this with Table VIII.) Elements in Group II have 2 outermost electrons, elements in Group III have 3 outermost electrons, and so on. The noble gas elements at the far right of the table have 8 outermost electrons. The transition elements may be thought of as arranged in **subgroups,** and all of these have 2 outermost electrons with the exception of the copper-silver-gold subgroup, which has only one outermost electron.

The horizontal rows of elements are called **periods.** All the elements in a given period have the same number of shells of electrons. For example, the elements in Period 1 have but one shell of electrons. Those in Period 2 have two shells,

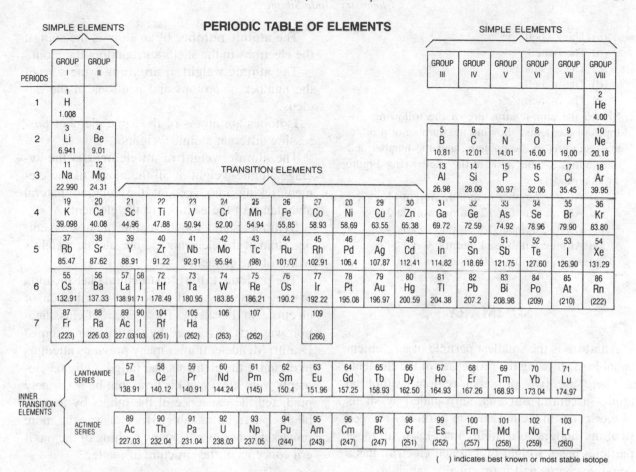

Fig. 12.

and so on. It is important to note that the last element of each period is a noble gas element. The lanthanide elements of the inner transition series are part of Period 6, and the actinide elements of the inner transition series are part of Period 7.

All four structural types of elements are shown in the table. The noble gas elements form a group at the extreme right. The simple elements are found in Groups I through VII, and since their logical conclusion, the completely filled shell, is found in the noble gases, they may be said to be found in Groups I through VIII. The transition elements are at the center. The lanthanide and actinide elements are extracted from the table and listed at the bottom.

Elements 106, 107, and 109 have not been named as yet. Element 106 was discovered in 1974 both at the University of California at Berkeley and at the Joint Institute for Nuclear Research at Dubna, U.S.S.R. Elements 107 and 109 were discovered in 1981 and 1982 respectively at the Institute for Heavy Ion Research at Darmstadt, Germany.

It has been pointed out that the chemical behavior of elements is based upon the electronic structure of their atoms, particularly the structure of the outer shell. Since each *group* of elements has the same structure in the outer shell, we can expect the members of a group to show similar chemical behavior. For this reason we can expect to find much use for the periodic table as we explore the chemical behavior of elements.

Problem Set No. 3

1. Using Equation (1) on p. 17, compute the maximum number of electrons possible in:

(a) The k-shell.
(b) The l-shell.
(c) The m-shell.
(d) The n-shell.
(e) The o-shell.

2. Sketch the atomic structure of the following isotopes of elements, showing the number of protons and neutrons in the nucleus, and the number and distribution of electrons in the shells: (Use Figure 11 as a model)
  (a) Atomic number 5; atomic weight 11.
  (b) Atomic number 12; atomic weight 24.
  (c) Atomic number 18; atomic weight 40.
  (d) Atomic number 34; atomic weight 79.
  (e) Atomic number 40; atomic weight 93.

SUMMARY

An **atom** is the smallest particle of an element capable of showing the properties of the element.

There are two principal regions within an atom—a central **nucleus**, surrounded by **shells of electrons.** In the nucleus are positively charged **protons** and neutral **neutrons.** Each of these particles has a weight of 1. Each **electron** has a negative charge and is practically weightless.

The **atomic number** of an atom is the sum of the electrons in the shells surrounding the atom.

The **atomic weight of an atom** is the sum of the number of protons and neutrons in the nucleus.

Isotopes are atoms of the same element possessing different atomic weights.

The **atomic weight of an element** is the average atomic weight of all the atoms of the element taking into account the relative natural abundance of the isotopes of the element.

Symbols stand for the name of an element, one atom of the element, and one atomic weight's worth of the element.

The **periodic table** is a structural classification of the elements based upon the distribution of electrons in the shells. This system of classification was first developed by the Russian chemist Dimitri Mendeleeff after many previous attempts to organize the elements had failed. Only 63 elements were known at his time, but he accurately predicted the existence of the others by leaving blank spaces in his table. His feat was the more remarkable because he knew nothing of the modern concepts of the structure of matter.

CHAPTER 4

COMPOUNDS

Elements in the free or uncombined state make up only a small fraction of matter. Most of matter occurs as compounds or mixtures of compounds. Let us now put our Bohr model of the atom to the test to see whether it is useful in explaining how elements can combine to form all the various compounds.

THE NOBLE GAS ELEMENTS

Take another look at Table VIII on page 18. Pay special attention to the elements which are just above the separating lines. Notice that except for helium, all these elements have eight electrons in the outermost shell. Then notice that the next element in each case has one electron in a new shell. Why doesn't this new electron go into one of the existing shells? The answer is that it simply doesn't fit, which is another way of saying that the shells of electrons in the elements just above the separating lines are, for the moment at least, filled up or saturated with electrons.

Now look up each of the elements just above the separating lines in the periodic table on page 21. You will find all of them in the column at the far right under the heading Group VIII, Simple Elements, which is the same as Group VIII, Noble Gas Elements. Each of them occurs at the end of a period. Among the noble gas elements, krypton, xenon, and radon react with reluctance to form compounds, and so far it has not been possible to react helium, neon, and argon.

You might ask why we should bring up this group of elements which form few compounds in a chapter devoted to the formation of compounds. Well, these elements possess a structure so stable that they resist compound formation. Other elements are less stable since they more readily form compounds. It is suggested that when active elements combine to form compounds, they undergo a rearrangement of their electronic structures in order to gain an electronic configuration similar to that of a noble gas element. Such a rearrangement causes the active elements to become structurally more stable.

VALENCE

The tendency of elements to form compounds through a shift of electronic structure is known as **valence.** Actually the term valence may be used to indicate two different things. One is va-

Table IX
Properties of Noble Gas Elements

Property	Helium He	Neon Ne	Argon Ar	Krypton Kr	Xenon Xe	Radon Rn
Atomic Number	2	10	18	36	54	86
Atomic Weight	4.003	20.179	39.948	83.80	131.29	222
Density, g./cc.	0.00018	0.0009	0.00179	0.00374	0.0058	0.0099
Solubility, ml./100 ml. water	1.49	1.5	5.6	6.0	28.4	0.000002
Boiling Point, °C.	−268.9	−246.0	−185.7	−152.3	−107.1	−61.8
Melting Point, °C.	272.2	−248.7	−189.2	−156.6	−111.9	−71.0

lence mechanism, that is, the manner in which elements attain a stable electronic distribution. The other is valence number, that is, the number of electrons of an element involved in forming a compound. Let us examine first the valence mechanism, the process by which compounds are formed.

ELECTROVALENCE

Consider for a moment the structure of an atom of sodium. It has one electron in its outermost shell and eight electrons in its next outermost shell. If the lone electron were to be removed from the sodium atom, the remaining electronic structure would be identical to the structure of neon, a noble gas element which immediately precedes sodium in the periodic table. The removal of the electron would change the nature of the particle by causing it to have one excess positive charge. It would no longer be a sodium atom for, although its nucleus is still that of sodium, it would possess an insufficient number of electrons to be a sodium atom. Nor would it be neon, for its nucleus has too many protons. An electrically charged particle of the type described is called an ion. The one being considered is a sodium ion. Ions possess properties which are totally different from the atoms from which they come.

The idea of forming an ion from an atom is reasonable enough, but where can the electron go? The elements near sodium in the periodic table, like potassium or calcium or magnesium, would not accept an additional electron, for it would not bring them nearer a stable electronic configuration. But over near the other end of the periodic table are elements like chlorine. Chlorine has seven electrons in its outermost shell. The addition of one more electron would give it the same stable configuration as argon, another noble gas element. The addition of the electron to the chlorine atom would form a particle with one excess negative charge. It would be neither a chlorine atom nor an argon atom. It would be a chloride ion. Once the electron has been transferred from the sodium atom to the chlorine atom, we then have oppositely charged ions which are

capable of attracting one another electrically. They do so to form the familiar compound sodium chloride, which is ordinary table salt. Figure 13 shows the formation of this compound. The process of forming a compound through the transfer of electrons is called electrovalence.

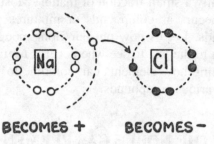

Fig. 13. Electrovalence

A careful consideration of the periodic table will lead to the discovery of the elements which show electrovalence. Elements in Groups I, II, and III give up 1, 2, or 3 electrons respectively to form positive ions. These elements are said to exhibit positive valence. Elements in Groups V, VI, and VII accept 3, 2, or 1 electron respectively to form negative ions. These elements are said to exhibit negative valence. Table X gives the symbols for typical ions formed by elements in each of these groups.

Theoretically, the elements in Group IV can form ions either by gaining or by losing electrons.

It is not difficult to see that it takes energy to remove an electron from a sodium atom, or to force an electron into a chlorine atom. Similarly it seems reasonable that it takes more energy to strip the 2 electrons from a magnesium atom than to remove the 1 electron from a sodium atom. The ease with which an element loses or gains electrons is a measure of its activity. On the basis of energy considerations we may state the following. Elements in Groups I and VII are more active than those in Groups III and V. Only elements near the bottom of Group IV form ions. Thus, it can be seen that activity decreases as we consider elements toward the center of the periodic table.

Within a given group, there is also a range of chemical activity. Considering Group I, the negative electron to be removed from the hydrogen

Table X						
Symbols of Typical Ions						
Group	*I*	*II*	*III*	*V*	*VI*	*VII*
Ionic Symbols	Na^+	Ca^{++}	Al^{+++}	N^{---}	O^{--}	F^-
	K^+	Mg^{++}			S^{--}	Cl^-

atom is much closer to the positive nucleus than the electron to be removed from the cesium atom. Thus it can be seen that it takes less energy to form a cesium ion than to form a hydrogen ion. Therefore, cesium is much more active than hydrogen. On the other hand, in Group VII similar reasoning tells us that it will require more energy to force an electron into a bromine atom, where the positive nucleus is buried within a cloud of negative electrons, than to force an electron into a fluorine atom, where the positive nucleus is relatively close to the outermost shell and can help attract the extra electron in. On the basis of energy considerations we may state that the most active electrovalent elements are to be found in the **lower left** and **upper right** areas of the periodic table.

The transition and inner transition elements also form ions. However, because both their outermost and second outermost shells are unfilled, they give up not only their outermost electrons to form ions, but they may also give up some electrons from their second outermost shell as well. Thus it is common to find these elements forming two or more different positive ions.

The compounds formed by electrovalence, then, really consist of oppositely charged ions packed and held together by electrical attraction. Such compounds are called **ionic agglomerates.**

COVALENCE

On the basis of electrovalence, we would expect an element like carbon, which is in Group IV, to be fairly inert and form few compounds. Yet this element forms more compounds than all the other elements put together. Obviously, then,

there must be some other valence mechanism.

Carbon has four electrons in its outermost shell. Hydrogen has one electron in its only shell. Suppose that four hydrogen atoms were to approach a carbon atom very closely, so closely that the shell of each hydrogen atom penetrated into the outermost shell of the carbon atom. The electrons in these interpenetrated shells would then be influenced by the nuclei of both types of atoms. Both atoms would, in a sense, be sharing electrons. What would be the net effect? Figure 14

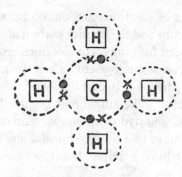

Fig. 14. Covalence

shows us. The electron of each hydrogen atom is indicated by an x, and the carbon electrons are indicated by dots (the inner carbon atoms are not shown). We can see that two electrons are now associated with each hydrogen atom giving them the stable helium configuration, and eight electrons are associated with the carbon atom giving it the stable neon configuration. Both types of atoms have attained stable structures through this sharing process. The compound described is methane, the principal ingredient of natural gas used in cooking. The process of forming a compound through the sharing of pairs of electrons is called **covalence.**

A pair of electrons shared between two atoms

is often called a **bond.** In methane, carbon is united to four hydrogen atoms by single bonds. Many compounds exist in which two or even three pairs of electrons are shared by two atoms. Figure 15 shows the bonding in carbon dioxide, a gas which bubbles out of carbonated water, and of acetylene, a gas commonly used in welding.

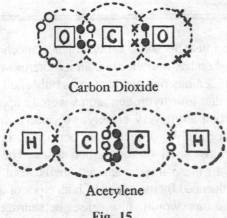

Carbon Dioxide

Acetylene

Fig. 15.

Two pairs of electrons are shared between each oxygen atom and the central carbon atom in carbon dioxide, forming eight electrons around each of the three atoms present. Three pairs of electrons are shared between the two carbon atoms in acetylene, and a single pair is shared between the carbon and hydrogen atoms. Carbon dioxide is said to have two **double bonds,** and acetylene is said to have a **triple bond** between its two carbon atoms.

The net effect of covalence is to form tiny particles of compounds containing a definite number of atoms. These discrete individual particles which possess all the properties of the compound are called **molecules.** Molecules are present only in covalent compounds. Electrovalent compounds do not have molecules, but rather are made up of ions packed together. Table XI summarizes the differences between electrovalence and covalence.

VALENCE NUMBER

The **valence number** of an element is the number of electrons of the element involved in

Table XI
Valence Mechanisms

Mechanism:	Electrovalence	Covalence
Process:	Complete transfer of electrons	Sharing of pairs of electrons
Via:	Formation of ions	Interpenetration of atoms
Product:	Ionic agglomerates	Molecules

the formation of a compound. Since free elements are not combined with other elements, **elements in the free state have a valence number of zero.** Most elements exhibit a variety of valence numbers depending upon the particular compound they happen to be part of. To help you determine the valence number of an element in a compound, the following general rules are given:

1. Elements of Group I of the periodic table normally have a valence number of $+1$.
2. Elements of Group II normally have a valence number of $+2$.
3. Elements of Group VII normally have a valence number of -1 in **binary compounds** (compounds which contain only 2 elements).
4. In electrovalent compounds in general:
   a. The valence number of an ion is numerically equal to the charge on the ion.
   b. Positive ions have positive valence numbers.
   c. Negative ions have negative valence numbers.
5. In covalent compounds in general:
   a. The valence number of an atom in a covalent compound is numerically equal to the number of its electrons shared with atoms different from itself. For example, referring again to Figure 15, the carbon atom in carbon dioxide shares all four of its outermost electrons with oxygen atoms, so its valence number is 4. But in acetylene, three carbon electrons are shared with another carbon atom while one electron is shared with a hydrogen atom.

The carbon-to-carbon bonds don't count, so the valence number of carbon in acetylene is 1.

b. Oxygen always has a valence number of -2 in its compounds (except peroxides, where its valence number is -1).

c. Elements like carbon, silicon, nitrogen, phosphorus, sulfur, and chlorine, when they are centrally located in covalent molecules, normally have positive valence numbers.

6. The net sum of all the valence numbers exhibited in a given compound must be zero.

These rules generally apply in most chemical compounds. Where exceptions occur, they will be pointed out.

FORMULAS

The **formula** of a compound is a ratio of the number of atoms of each element present in the compound. In electrovalent compounds, the formula gives the simplest ratio of constituents in whole numbers. In covalent compounds, the formula gives the exact number of atoms of each element present in a molecule of the compound.

The symbols of the elements present are used in writing formulas. If more than one atom of an element is required in the formula, a subscript numeral is written behind the symbol of the element to indicate the number of its atoms in the formula. For example, the formula for water is H_2O. This means that in every molecule of water there are two atoms of hydrogen and one atom of oxygen. Note that when only one atom of an element is present in the formula, the subscript 1 is understood and not written. The formula for sodium chloride, NaCl, tells us that this compound contains equal numbers of sodium and chlorine atoms. We know from previous discussions that in this compound the atoms are actually present as ions. A formula gives no indication as to whether a compound is electrovalent or covalent. This characteristic must be ascertained from the properties of the compound.

FORMULAS AND VALENCE

If we know the valence of each element in a compound, we can easily write its formula. Let us look at a few examples.

EXAMPLE 1: A compound consists solely of magnesium and chlorine. What is its formula?

SOLUTION: Mg, a Group II element, has a valence number of $+2$. Cl, a Group VII element, has a valence number of -1.

Therefore, to form a compound in which the sum of the valence numbers is zero, it will take two chlorine atoms to nullify each magnesium atom. The formula of this compound must therefore be:

$$MgCl_2.$$

The name of this compound is magnesium chloride. The suffix **-ide** is used with the root of the name of the negative element in binary compounds. The terms *oxide, sulfide, nitride, phosphide, carbide, fluoride, bromide,* and *iodide* appear in the names of compounds in which these negative elements are combined with one other positive element to form a binary compound.

Look carefully at the formula of magnesium chloride, $MgCl_2$. Behind Mg, the subscript 1 is understood. Behind Cl is the subscript 2. Do you see that the valence number of each element has been crisscrossed and written as a subscript behind the symbol of the other element? Let us try this idea with another example.

EXAMPLE 2: What is the formula of aluminum oxide?

SOLUTION: Al, a Group III element, has a valence number of $+3$. O, a Group VI element, has a valence number of -2.

Crisscrossing the valence numbers and using them as subscripts, we have the formula:

$$Al_2O_3.$$

Note that the sign of the valence numbers is ignored when we write formulas.

Does this formula for aluminum oxide satisfy the rule of zero net valence for compounds? Let's check it.

For Al: $2 \times (+3) = +6.$
For O: $3 \times (-2) = -6.$
Net valence (sum) $= \quad 0.$

EXAMPLE 3: What is the formula of calcium sulfide?

SOLUTION: Ca, a Group II element, has a valence number of $+2$. S, a Group VI element, has a valence number of -2. Crisscrossing the valence numbers and writing them as subscripts, we have the formula:

$$Ca_2S_2.$$

However, this is not the simplest formula for this compound. This formula tells us that the ratio of calcium to sulfur atoms is 2:2. This, of course, is the same as a ratio of 1:1. Therefore, to write this formula in its simplest form, we reduce the subscripts to 1, and the formula becomes:

$$CaS.$$

Now let's look at the relationship between formulas and valence the other way. Suppose we are given the formula of a compound and we have to find the valence numbers of the elements present. Let's look at some examples.

EXAMPLE 4: What are the valence numbers of the elements in sulfur dioxide, SO_2?

SOLUTION: Our rules on p. 26 tell us that oxygen has a valence number of -2. Since there are 2 oxygen atoms in our formula, the total negative valence is then -4. Therefore, to satisfy the rule of zero valence in the compound, the valence number of S in SO_2 must be $+4$.

EXAMPLE 5: What are the valence numbers of the elements in sulfuric acid, H_2SO_4?

SOLUTION: O always has a valence number of -2. H, a Group I element, has a valence number of $+1$. The 4 O atoms give us a negative valence of $4 \times (-2) = -8$. The 2 H atoms give us a positive valence of $2 \times (+1) = +2$. Therefore, in order to make the net valence of the compound zero, S in H_2SO_4 has a valence number of $+6$.

RADICALS

In many chemical compounds there are clusters of elements which behave as if they were a single element. Such a group of elements is known as a **radical.** Consider the following series of compounds.

Series A

Sodium chloride	NaCl
Sodium hydroxide	NaOH
Sodium nitrate	$NaNO_3$

Series B

Sodium sulfide	Na_2S
Sodium sulfate	Na_2SO_4
Sodium carbonate	Na_2CO_3

In series A, the hydroxide (OH) and the nitrate (NO_3) groups have behaved toward sodium in exactly the same way as a single chlorine atom. Similarly, in series B, the sulfate (SO_4) and carbonate (CO_3) groups have behaved toward sodium in exactly the same way as a single sulfur atom. All of these groups are radicals.

The atoms within a radical are held together by covalent bonds, but in each case, they contain either an excess or a deficiency of electrons, causing the radical to possess an electrical charge. Thus, radicals are really complex ions. They then combine as a unit with other ions to form electrovalent compounds.

Radicals possess a net valence number equal in magnitude and sign to the net charge on the radical, just like any other ion. Table XII gives the names, formulas, and valence numbers of the common radicals.

Table XII
Radicals

Valence Number $+1$	*Valence Number* -2
Ammonium NH_4^+	Carbonate CO_3^{--}
	Chromate CrO_4^{--}
Valence Number -1	Dichromate $Cr_2O_7^{--}$
Acetate $C_2H_3O_2^-$	Sulfate SO_4^{--}
Bicarbonate HCO_3^-	Sulfite SO_3^{--}
Chlorate ClO_3^-	
Hydroxide OH^-	*Valence Number* -3
Cyanide CN^-	Phosphate PO_4^{---}
Nitrate NO_3^-	
Nitrite NO_2^-	
Permanganate MnO_4^-	

Since ammonium is a positive radical, it will form compounds with all the negative radicals. Notice how the formulas of these compounds are written.

Ammonium acetate	$NH_4C_2H_3O_2$
Ammonium carbonate	$(NH_4)_2CO_3$
Ammonium phosphate	$(NH_4)_3PO_4$

Note that it takes two ammonium ions to satisfy

the valence of the carbonate ion, and three ammonium ions to satisfy the valence of the phosphate ion. (Remember ammonium phosphate from the matchstick?)

Look carefully at the names and formulas of the radicals. The suffixes **-ite** and **-ate** occur repeatedly. The suffixes are used only with radicals containing oxygen atoms. Notice that "-ite" radicals contain less oxygen than "-ate" radicals. For example:

Sulfite, SO_3	Sulfate, SO_4
Nitrite, NO_2	Nitrate, NO_3

Note also that there is no definite number of oxygen atoms in either type. The formulas of each radical must be learned individually through repeated use.

FORMULA OR MOLECULAR WEIGHTS

Just as symbols represent more than just the name of an element, so formulas stand for more than merely the name of a compound. A formula stands for three things:

1. The name of a compound.
2. One molecule of the compound (if it is covalently bonded).
3. A quantity of the compound equal in weight to its **formula weight.**

This concept of the formula weight of a compound is one of the most important ideas in chemistry. Its definition is very simple. **The formula weight of a compound is the sum of all the atomic weights of the elements present in the formula of the compound.** The formula weight of sodium chloride, NaCl, is found as follows:

Atomic weight of Na	22.990
Atomic weight of Cl	35.453
Formula weight of NaCl	58.443

Similarly, the formula weight of water, H_2O, would be found thus:

Atomic weight of H ($\times 2$)	2.016
Atomic weight of O	15.999
Formula weight of H_2O	18.015

Since the formula of a covalent compound represents the constituents of a molecule of the compound, the formula weight is usually referred to as the **molecular weight** of the compound. As a matter of fact, since a formula gives no indication as to the type of bonding present in a compound, the term **molecular weight** is commonly used even with electrovalent compounds, even though no molecule is present in these compounds. Thus, in usage the terms formula weight and molecular weight are completely interchangeable.

A quantity of a compound equal in weight to its formula weight is called a **mole.** For example, 18.015 units of weight of water is one mole of water. 18.015 grams of water would be one **gram-mole;** 18.015 pounds of water would be one **pound-mole;** etc. Any quantity of a given compound can be expressed in terms of the number of moles of the compound present. The number of moles of a compound is found by using the following expression:

$$\frac{\textbf{Actual weight}}{\textbf{Formula weight}} = \textbf{Number of moles.} \quad (1)$$

We will begin to see in the next chapter how fundamentally important the concept of the mole is in the science of chemistry.

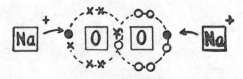

Fig. 16.

Problem Set No. 4

1. Figure 16 is a diagram of the electronic configuration in a compound (only the outermost electrons of each atom are shown). Determine the valence number of each atom in the compound. What is the name of the compound?
2. A compound consists solely of aluminum and sulfur. Work out its formula. What is its name?
3. A compound consists solely of magnesium and nitrogen. Work out its formula. What is its name?
4. What are the valence numbers of each atom in sodium carbonate, Na_2CO_3?
5. What are the valence numbers of each atom in calcium nitrate, $Ca(NO_3)_2$?

6. What is the formula of ammonium sulfate?
7. What is the formula of magnesium phosphate?
8. Compute the formula weight of sugar, $C_{12}H_{22}O_{11}$.
9. Compute the formula weight of aluminum sulfate, $Al_2(SO_4)_3$.
10. 200 grams of sodium hydroxide, NaOH, are to be used in making up a solution. How many gram-moles would that be?

SUMMARY

Noble gas elements resist formation of compounds because of the great stability of the electronic configurations in their atoms. In forming compounds, elements have their electronic distribution rearranged to match that of the noble gas elements.

The rearrangement of electrons is a process known as **valence mechanism.** Compound formation through the process of **transfer of electrons** from one atom to another is called **electrovalence.** The **ions** produced by this transfer are packed together electrically to form **ionic agglomerates.** Compound formation through the process of **sharing electrons** between atoms is called **covalence.** Discrete particles called **molecules** are produced by this process.

The **valence number** of an element is the number of its electrons involved in the formation of a compound. Valence numbers may be **positive** or **negative,** but the **net sum** of the valence numbers of elements in a compound **must be zero.**

A **formula** is a ratio of the number of atoms of each element present in a compound. It contains the **symbols** of each element present, with **subscripts** behind each symbol to indicate the number of atoms of the element present.

A **radical** is a group of elements which behaves as a single element. All radicals are ions, possessing an electrical charge.

A **mole** is a quantity of a compound equal in weight to the **formula weight,** or **molecular weight,** of the compound. The **formula weight** is the sum of the atomic weights of the elements present.

CHAPTER 5

LAWS OF CHEMISTRY

We have seen how our model of an atom has given us a reasonable explanation of how atoms are combined in compounds. Now we want to look at some of the basic laws of chemistry. These laws were discovered only after years of pains-taking observation of the behavior of nature. It should be kept in mind that *they were all known before our atomic model was created.* Each contributed to the development of our model. However, our primary concern now with these laws is with the assistance they can give us in understanding chemical change.

CONSERVATION OF MATTER

The Law of Conservation of Matter states that matter is neither created nor destroyed during chemical change. This means that the sum of the weights (total mass) of the substances entering a chemical change must be precisely equal to the sum of the weights (total mass) of the substance formed as a result of chemical change. This law has been verified by repeated study of chemical changes using delicate balances to measure the weights (mass) of **reactants** and **products.** Actually mass and energy are related, and it is more accurate to say that mass-energy is neither created nor destroyed during chemical change. We shall discuss this topic in Chapter 23 on nuclear chemistry.

As we have seen, chemical change involves a **redistribution** of electrons, either by transfer or by sharing, but no new electrons are formed in chemical change, nor are any destroyed. The nuclei of atoms, which possess all the weight, remain unchanged and are carried along into new combinations solely as a result of the redistribution of electrons. Thus, our atomic model is consistent with the Law of Conservation of Matter.

DEFINITE PROPORTIONS

The Law of Definite Proportions states that a given compound always contains the same elements combined in the same proportions by weight. The decomposition of a compound into its elements for the purpose of finding out how much of each element is present is known as **analysis** of the compound. Repeated analysis of a compound always shows that it contains the same elements in the same weight proportions.

For example, water always contains 8 parts by weight of oxygen to 1 part by weight of hydrogen. Let us see if these results are consistent with our concepts of atomic structure and compound formation. Oxygen, with atomic weight 16.0, has 6 electrons in its outermost shell. Hydrogen, with atomic weight 1.0, has 1 electron in its shell. Oxygen needs 2 electrons to fill its outermost shell. Our concept of compound formation tells us that 2 hydrogen atoms are required to provide sufficient electrons to fill the outer shell of oxygen. Furthermore, this gives us a weight proportion of 16 parts by weight of oxygen (1 atom) to 2 parts by weight of hydrogen (2 atoms), which is consistent with the 8 to 1 proportion always found in the analysis of water.

The Law of Definite Proportions has further significance. The process of causing elements to combine to form compounds is known as **synthesis** of compounds. The Law of Definite Proportions dictates that a compound formed by synthesis must contain the same weight (mass) proportions of its elements as any other samples of this compound. Thus, water produced in a laboratory by combining oxygen and hydrogen must contain 8 parts by weight of oxygen to 1 part by weight of hydrogen, the same as any other water sample. Now suppose that one took 8 parts by weight of oxygen and 2 parts by weight of hydrogen and attempted to combine them. What would happen? Well, it can be seen that there is too much hydrogen. The 8 parts of oxygen would combine with 1 part of hydrogen, and the rest of the hydrogen would remain unchanged. In this case the oxygen is said to be the **limiting reactant,** for the amount of water formed is based upon the amount of oxygen present. Likewise, in this case, there is said to be an **excess** of hydrogen present, for there is more present than oxygen can combine with.

In similar fashion, if one were to begin with 10 parts by weight of oxygen and 1 part by weight of hydrogen, 8 parts of oxygen would combine with the 1 part of hydrogen to form water, and the rest of the oxygen would be left unchanged. Here, the hydrogen is the limiting reactant and an excess of oxygen is present. This concept of limiting and excess reactants is very important, for in all chemical changes that involve two or more reactants, one of the reactants will always be the limiting reactant, and the others will be in excess.

AVOGADRO'S HYPOTHESIS

Avogadro's Hypothesis states that equal volumes of gases measured at the same temperature and pressure contain equal numbers of molecules. All gases exist as molecules. By finding the ratio of weights of equal volumes of various gases, we can find the ratio of their molecular weights. For example, let us consider again the compound water, and its elements hydrogen and oxygen. We can easily convert water to a gas (steam) and weigh a given volume of it. Likewise the same volume of hydrogen and oxygen, both gases, can be brought to the same temperature as the steam and weighed. The weight ratios found by this procedure always turn out to be as follows:

Hydrogen	1 part by weight
Oxygen	16 parts by weight
Water	9 parts by weight

The sample of oxygen weighs 16 times as much as the sample of hydrogen, and the sample of steam weighs 9 times as much as the hydrogen.

Since, by Avogadro's Hypothesis, each of these samples contains the same number of molecules, the individual molecules of each of these substances must possess these same weight proportions. Now we know the weight of one of these molecules. The formula of water is H_2O, and its molecular weight, obtained by adding the atomic weights in the formula, is 18. Recalculating our weight ratio found above to a basis of 18 for water we get:

Hydrogen	2 parts by weight
Oxygen	32 parts by weight
Water	18 parts by weight

Since this ratio is a ratio of molecular weights, and since the actual molecular weight of water is 18, the actual molecular weight of hydrogen must be 2, and the actual molecular weight of oxygen must be 32.

Therefore the molecule of hydrogen gas must contain 2 atoms of hydrogen, because the atomic weight of hydrogen is 1. Similarly the molecule of oxygen must contain 2 atoms of oxygen, because the atomic weight of oxygen is 16, one half of the molecular weight. The formula of hydrogen gas is therefore written H_2 to indicate the 2 atoms in the molecule. The formula for oxygen gas is O_2. Both of these molecules are covalently bonded.

Avogadro's Hypothesis is thus very useful in finding the molecular weight and formula of a gaseous substance, provided that its weight can be compared with the weight of a substance whose formula is known. Experimental and mathematical studies of Avogadro's Hypothesis have indicated its accuracy beyond reasonable doubt.

EQUATIONS

An equation is simply a statement of a chemical change using chemical symbols. When sulfur or any other substance burns in air, it is combining with oxygen in air to produce an oxide. Let us look at this reaction in the form of a chemical equation.

$$S \quad + \quad O_2 \quad = \quad SO_2$$
Sulfur   *Oxygen*   *Sulfur dioxide*

Examine the equation closely. Is it consistent with the Law of Conservation of Matter? In other words, are there equal numbers of each type of atom on each side of the equation? Yes, we see that this is so. This equation therefore is said to be **balanced.** An equation is meaningless unless it is balanced.

This equation tells us more than merely that sulfur combines with oxygen to produce sulfur dioxide. It has quantitative significance just as symbols themselves do. It tells us that one atomic weight's worth of sulfur reacts with one molecular weight's worth of oxygen to produce one molecular weight's worth of sulfur dioxide. If units of grams are used, this would be:

$$S \quad + \quad O_2 \quad = \quad SO_2$$
$$32.1 \text{ g.} \qquad 32 \text{ g.} \qquad 64.1 \text{ g.}$$

In other words, this equation tells us that one mole of sulfur combines with one mole of oxygen to produce one mole of sulfur dioxide.

Let us look at another reaction. You will recall that when copper is heated in air, black copper oxide is formed. This reaction is indicated as follows:

$$\underset{\textit{Copper}}{Cu} \quad + \quad \underset{\textit{Oxygen}}{O_2} \quad = \quad \underset{\textit{Copper oxide}}{CuO}$$

What about our Law of Conservation of Matter now? Do you see that we have apparently destroyed some oxygen? This equation is not balanced. It is called a **skeleton equation,** for it indicates only the names of the substances involved.

This equation would be balanced if we could put a subscript 2 behind the O of CuO to make it CuO_2. But this would violate the Law of Definite Proportions, because black copper oxide always has the formula CuO. In balancing equations, **the subscript in the formulas of compounds may not be changed.**

A skeleton equation is balanced by placing numbers, called **coefficients,** in front of the formulas of the substances in the reaction. Look again at our skeleton equation. An even number of oxygen atoms appears on the left side of the equation. By placing the coefficient 2 in front of CuO, we would have two oxygen atoms on each side of the equation, for the coefficient multiplies

all the symbols in the formula immediately behind it. This would change our equation to read:

$$Cu + O_2 = 2 \, CuO.$$

Now we have too much copper on the right. This can be remedied by placing another coefficient 2 in front of the Cu, giving us the following.

$$2 \, Cu + O_2 = 2 \, CuO.$$

Now the equation is balanced. We have 2 copper atoms and 2 oxygen atoms on each side of the equation. The balanced equation now reads 2 moles of copper combine with 1 mole of oxygen to produce 2 moles of copper oxide. The following expression shows how the weights of each of the substances in the balanced equation may be indicated:

$$2 \, Cu + O_2 = 2 \, CuO$$
$$2 \times 63.5 \quad 32 \quad 2(63.5 + 16)$$
$$127.0 + 32 = 159.0$$

So, 127.0 units of weight of copper combine with 32 units of weight of oxygen to form 159.0 units of weight of copper oxide. These units of weight may be grams, pounds, tons, etc., just so long as all three weights are expressed in the same units. This weight relationship also tells us that copper and oxygen combine in a weight ratio of 127.0 parts by weight of copper to 32 parts by weight of oxygen. Similarly, 159.0 parts by weight of copper oxide are formed for every 32 parts by weight of oxygen or every 127.0 parts by weight of copper.

Let us look at one more example. Butane gas (C_4H_{10}) is commonly used as a bottled gas in rural areas. It burns with oxygen to form carbon dioxide and water. The skeleton equation is:

$$C_4H_{10} + O_2 = CO_2 + H_2O.$$

Let us balance this skeleton equation using the "even numbers" technique described in the previous example.

1. Starting with oxygen, we see an even number of oxygen atoms on the left, and an odd number on the right. The CO_2 has an even number of oxygen atoms, so we have to work with the H_2O. Let's try a coefficient of 2. This would give us:

$$C_4H_{10} + O_2 = CO_2 + 2 \, H_2O.$$

This gives us an even number of oxygen atoms, but we need 10 hydrogen atoms and this gives us only 4 (2×2). Therefore we need a larger coefficient.

2. A coefficient of 5 would give us the right amount of hydrogen, but 5 is an odd number, so we must go to the next even multiple of 5 which is 10. This will do, but it gives us 20 hydrogen atoms on the right. By placing another coefficient of 2 in front of the C_4H_{10} we would also have 20 hydrogen atoms on the left. This gives us:

$$2\ C_4H_{10} + O_2 = CO_2 + 10\ H_2O.$$

Now our hydrogen is balanced and we have an even number of oxygen atoms on each side.

3. Now we look at the carbon. We have 8 carbon atoms on the left, so we need a coefficient of 8 in front of the CO_2 to balance the carbon. This gives us:

$$2\ C_4H_{10} + O_2 = 8\ CO_2 + 10\ H_2O.$$

We still have an even number of oxygen atoms on each side.

4. Now we are finally ready to balance the oxygen. There is a total of 26 oxygen atoms on the right side of the equation. A coefficient of 13 in front of the O_2 will give us 26 oxygen atoms on the left side. Now our equation is balanced and looks like this:

$$2\ C_4H_{10} + 13\ O_2 = 8\ CO_2 + 10\ H_2O.$$

This equation reads: 2 moles of butane combine with 13 moles of oxygen to produce 8 moles of carbon dioxide and 10 moles of water. The weight proportions involved are:

Reactants:
- Butane: $2(48 + 10) = 116$
- Oxygen: $13(32) = 416$ 532

Products:
- Carbon Dioxide: $8(12 + 32) = 352$
- Water: $10(2 + 16) = 180$ 532

The characteristics of a balanced equation may be summarized as follows:

1. It obeys the Law of Conservation of Matter.
2. It obeys the Law of Definite Proportions.

3. Its coefficients give the molar proportions of reactants and products involved in the reaction.

Symbols, formulas, and equations all have definite quantitative meanings. We are now ready to look at some numerical applications based upon these ideas.

PERCENTAGE COMPOSITION

If we know the formula of a compound, we can easily find the percentage by weight of each element present. A statement of the percentage of each element present in a compound is called its **percentage composition.** In chemistry, this composition is always on a weight basis unless specifically stated otherwise. Sometimes the composition of mixtures of gases is given on a volumetric basis.

The computation of percentage composition from the formula of a compound is based upon the meaning of symbols and formulas. Each symbol stands for one atomic weight's worth of the element it represents, and each formula stands for one molecular weight's worth of the compound it represents. Let us see how percentage composition calculations are carried out.

EXAMPLE 1: What is the percentage composition of water, H_2O?

SOLUTION:

	No. of Atoms	Atomic Weight	Total Weight
Hydrogen:	2	1.0	2.0
Oxygen:	1	16.0	16.0
Molecular weight of H_2O:			18.0

$$\text{Percentage of hydrogen} = \frac{2.0}{18.0} \times 100 = 11.1\%$$

$$\text{Percentage of oxygen} = \frac{16.0}{18.0} \times 100 = 88.9\%.$$

Note that the percentage of each element is found from the expression:

$$\frac{\textbf{Total wt. of element present}}{\textbf{Molecular wt. of compound}} = \textbf{\% of element.}$$

EXAMPLE 2: What is the percentage composition of sulfuric acid, H_2SO_4?

SOLUTION:

	No. of Atoms	Atomic Weight	Total Weight
Hydrogen:	2	1.0	2.0
Sulfur:	1	32.1	32.1
Oxygen:	4	16.0	64.0
Molecular weight of H_2SO_4:			98.1

Percentage of hydrogen $= \dfrac{2.0}{98.1} \times 100 = 2.0\%$

Percentage of sulfur $= \dfrac{32.1}{98.1} \times 100 = 32.7\%$

Percentage of oxygen $= \dfrac{64.0}{98.1} \times 100 = 65.3\%$.

EXAMPLE 3: Find the percentage of oxygen in calcium nitrate, $Ca(NO_3)_2$.

SOLUTION:

	No. of Atoms	Atomic Weight	Total Weight
Calcium:	1	40.1	40.1
Nitrogen:	2	14.0	28.0
Oxygen:	6	16.0	96.0
Molecular weight of $Ca(NO_3)_2$:			164.1

Percentage of oxygen $= \dfrac{96.0}{164.1} \times 100 = 58.5\%$.

Note particularly how the number of atoms of each element was obtained.

EXAMPLE 4: An iron ore field contains ferric oxide, Fe_2O_3, also known as **hematite,** mixed with rock which bears no iron. Naturally, both hematite and rock are scooped up in the giant shovels used in mining the ore. Samples taken at various spots in the ore field show that the field contains 80% hematite and 20% rock. Find the weight of pure iron in one ton of this ore, and the percentage of iron in the ore field.

SOLUTION: (1) Wt. of Fe_2O_3 per ton of ore:

2,000 × 0.80 = 1600 lbs. of Fe_2O_3 per ton of ore.

(2) Percentage of Fe in Fe_2O_3:

	No. of Atoms	Atomic Weight	Total Weight
Iron:	2	55.8	111.6
Oxygen:	3	16.0	48.0
Molecular weight of Fe_2O_3:			159.6

Percentage of Fe $= \dfrac{111.6}{159.6} \times 100 = 69.8\%$

(3) Wt. of Fe per ton of ore:

1600 × 0.698 = 1117 lbs. of Fe per ton of ore.

(4) Percentage of Fe in the field:

$\dfrac{1117}{2000} \times 100 = 55.9\%$ Fe in the ore field.

Example 4 shows how percentage composition problems may be a part of many different varieties of practical problems. Such fields as analytical chemistry, metallurgy, mining, minerology, and geology all make use of calculations of this type.

COMPUTATION OF FORMULAS

If we know the percentage composition of a compound, we can compute the **simplest formula** of the compound. As we have seen, a formula is a ratio of the number of atoms of each element present in the compound. The simplest formula gives this atomic ratio in terms of the smallest whole numbers of each type of atom present. For example, the true formula of hydrogen peroxide is H_2O_2. Its simplest formula would be HO. In general, the simplest formula is the true formula for all electrovalent compounds. In covalent compounds, where the formula represents the composition of the molecule of the compound, the true formula is either the same as the simplest formula, or it is some whole number multiple of it. We will learn how to calculate true formulas later, but for now let us concentrate on finding the simplest formula of a compound.

EXAMPLE 5: A compound is analyzed and found to contain 75% carbon and 25% hydrogen. Find its simplest formula.

SOLUTION: Since each different type of atom contributes to the total weight of the compound **in parcels of weight** equal to its own atomic weight, we can divide the weight percent of a given element by its atomic weight to get the relative number of atoms of the element contributing to the total weight percent. For the compound under consideration this would be:

Carbon: $\dfrac{75}{12} = 6.25$

Hydrogen: $\dfrac{25}{1} = 25.0$

Thus we have 6.25 carbon atoms for every 25 hydrogen atoms in this compound. To reduce these numbers to the simplest whole numbers, we divide each by the smaller. The entire calculation would then be as follows:

$$\text{Carbon:} \quad \frac{75}{12} = 6.25; \quad \frac{6.25}{6.25} = 1$$

$$\text{Hydrogen:} \quad \frac{25}{1} = 25; \quad \frac{25}{6.25} = 4.$$

Therefore the simplest formula of this compound is CH_4.

EXAMPLE 6: A compound contains 21.6% sodium, 33.3% chlorine, and 45.1% oxygen. Find its simplest formula.

SOLUTION:

$$\text{Sodium:} \quad \frac{21.6}{23.0} = 0.94; \quad \frac{0.94}{0.94} = 1$$

$$\text{Chlorine:} \quad \frac{33.3}{35.5} = 0.94; \quad \frac{0.94}{0.94} = 1$$

$$\text{Oxygen:} \quad \frac{45.1}{16.0} = 2.82; \quad \frac{2.82}{0.94} = 3.$$

Therefore the formula of this compound is $NaClO_3$.

EXAMPLE 7: Some crystalline solids have molecules of water forming part of their crystal structure. Such solids are known as **hydrates.** Ordinary household washing soda, made up of sodium carbonate and water, is a typical hydrate. The percentage of water present can be found by measuring the loss in weight of a hydrate sample dried in a hot oven. A 20.00 g. sample of washing soda is dried in an oven. After drying it is found to weigh 7.57 g. Compute:
(a) The percentage of water in washing soda.
(b) The formula of washing soda.

SOLUTION:

(a) Percentage $H_2O = \dfrac{\text{loss in wt.}}{\text{original wt.}} \times 100 =$

$$\frac{20.00 - 7.57}{20.00} = 62.15\%.$$

(b) The percentage of Na_2CO_3 is $100\% - 62.15\%$
     $= 37.85\%$

$$Na_2CO_3: \quad \frac{37.85}{106} = 0.357; \quad \frac{0.357}{0.357} = 1$$

$$H_2O: \quad \frac{62.15}{18.0} = 3.45; \quad \frac{3.45}{0.357} = 10.$$

       (to the nearest whole number)

Therefore the formula of washing soda must indicate 1 part of Na_2CO_3 and 10 parts of H_2O. Its formula is

written as follows: $Na_2CO_3 \cdot 10\, H_2O$. This is the standard method of writing the formula of a hydrate. It indicates that the crystal contains 10 moles of water for every mole of sodium carbonate. Note particularly that since a molar ratio of constituents was sought, the molecular weights of each constituent were used in finding the molar ratio.

WEIGHT OR MASS RELATIONSHIPS IN EQUATIONS

We have seen that chemical equations tell us the number of moles of each substance involved in a given reaction. For example, the equation for the rusting of iron:

$$4\, Fe + 3\, O_2 = 2\, Fe_2O_3$$

tells us that iron combines with oxygen in a ratio of 4 moles of iron to 3 moles of oxygen, and that 2 moles of iron oxide are produced for every 4 moles of iron entering the reaction.

These molar ratios, in turn, indicate the ratio of weights of each substance involved. The equation tells us that iron and oxygen combine in a ratio of (4×55.8) parts by weight of iron to (3×32) parts by weight of oxygen, and that (2×159.8) parts by weight of iron oxide are thereby produced in this reaction. When we multiply the coefficient of a substance in a balanced equation by the formula weight of the substance, we obtain a quantity known as the **equation weight** of the substance. **The actual weight of substances involved in a chemical reaction are in the same ratio as their equation weights.**

Therefore if we know the balanced equation for a reaction and the actual weight of any one substance involved in the reaction, we can find the actual weight of any other substance participating in the reaction from the following proportion:

$$\frac{\textbf{Actual wt. of one substance}}{\textbf{Its equation weight}} = \frac{\textbf{Unknown actual weight}}{\textbf{Its equation weight}}$$

Let us look at an example involving the finding of actual weights.

EXAMPLE 8: 27.95 g. of iron are oxidized completely.
(a) What weight of oxygen combined with the iron?
(b) What weight of iron oxide was produced?

SOLUTION: First we write the balanced equation for the reaction and place the equation weight of each substance involved below its formula as follows:

$$4\ Fe\ +\ 3\ O_2\ =\ 2\ Fe_2O_3$$
$$4 \times 55.8 \qquad 3 \times 32 \qquad 2 \times 159.6$$

Part a: Substituting in the expression above to find the actual weight of oxygen we have: (Let x represent the unknown wt.)

$$\frac{27.95}{223.2} = \frac{x}{96}$$

$$x = \frac{27.95 \times 96}{223.2}$$

$$x = 12.0 \text{ g. of oxygen.}$$

Part b: Substituting in the expression above to find the actual weight of iron oxide we have:

$$\frac{27.95}{223.2} = \frac{x}{319.2}$$

$$x = \frac{27.95 \times 319.2}{223.2}$$

$$x = 39.97 \text{ g. of iron oxide.}$$

The steps, then, in solving this type of problem are:

1. Write the **balanced** equation for the reaction.
2. Find the equation weights of the substances concerned.
3. Equate the ratios of actual weights to equation weights for each of the substances, and solve for the unknown actual weight.

EXAMPLE 9: Sodium hydroxide, NaOH, may be prepared by treating sodium carbonate, Na_2CO_3, with calcium hydroxide, $Ca(OH)_2$, according to the following equation (skeleton):

$$Na_2CO_3 + Ca(OH)_2 = NaOH + CaCO_3.$$

What weight of NaOH can be produced from 74.2 g. of Na_2CO_3?

SOLUTION: Balanced equation:

$$Na_2CO_3 + Ca(OH)_2 = 2\ NaOH + CaCO_3$$
$$1 \times 106 \qquad\qquad 2 \times 40$$

Therefore:

$$\frac{74.2}{106} = \frac{x}{80}$$

$$x = \frac{74.2 \times 80}{106}$$

$$x = 56 \text{ g. of NaOH.}$$

Notice that it is assumed that there is sufficient calcium hydroxide present to react with all of the sodium carbonate. If any excess calcium hydroxide is used, it will remain unchanged, for the sodium carbonate is the limiting reactant in this case.

Problem Set No. 5

1. The formula of hydrogen gas is known to be H_2. Equal volumes of hydrogen and nitrogen gas are weighed at the same temperature and pressure. The nitrogen gas weighs 14 times as much as the hydrogen. What is the formula of nitrogen gas?
2. Find the percentage composition of sugar, $C_{12}H_{22}O_{11}$.
3. Find the percentage of copper in hydrated copper sulfate, $CuSO_4 \cdot 5\ H_2O$.
4. A sample of impure NaCl is found to contain 58% Cl. What is the percent purity of the sample?
5. A compound contains 52.9% Al and 47.1% O. Find its formula.
6. 50.88 g. of copper combine with 12.84 g. of sulfur. Find the formula of the compound formed.
7. Balance the following equations:

    (a) $NaCl + H_2SO_4 = Na_2SO_4 + HCl$.
    (b) $NH_3 + O_2 = N_2 + H_2O$.
    (c) $ZnS + O_2 = ZnO + SO_2$.
    (d) $C_3H_8 + O_2 = CO_2 + H_2O$.
    (e) $Ca_3(PO_4)_2 + SiO_2 + C = CaSiO_3 + CO + P$.

8. Lime, CaO, is prepared commercially by heating limestone, $CaCO_3$. The equation is:

    $$CaCO_3 = CaO + CO_2.$$

    What weight of lime could be obtained by heating 500 lbs. of limestone?

9. Potassium nitrate, KNO_3, decomposes when heated according to the following skeleton equation:

    $$KNO_3 = KNO_2 + O_2.$$

    (a) Balance the equation.
    (b) How many moles of O_2 will be formed from 12 moles of KNO_3?
    (c) What is the name of KNO_2? (See Table XII, p. 28)
    (d) What weight of KNO_2 will be formed by heating 12 gram-moles of KNO_3?

10. Nitrogen and hydrogen combine in the presence

of a catalyst to form ammonia, NH_3, according to the equation:

$$N_2 + 3 H_2 = 2 NH_3.$$

280 g. of N_2 and 100 g. of H_2 are admitted to a reaction chamber.

(a) Which of the two substances is the limiting reactant?

(b) How many moles of limiting reactant are present?

(c) How many moles of excess reactant are in excess?

(d) How many moles of NH_3 can be produced?

(e) What weight of NH_3 can be produced?

SUMMARY

Law of Conservation of Matter: Matter is neither created nor destroyed during ordinary chemical change.

Law of Definite Proportions: A given compound always contains the same elements combined in the same proportions by weight.

Avogadro's Hypothesis: Equal volumes of gases measured at the same temperature and pressure contain the same number of molecules.

An **equation** is a statement of a chemical change using chemical symbols. It must be **balanced** with **coefficients** so that it obeys the Laws of Conservation of Matter and Definite Proportions.

Coefficients give the **molar proportions** of **reactants** and **products** in chemical reactions.

Percentage composition is the **weight percent** of each element present in a compound.

The **simplest formula** of a compound gives the ratio of the number of atoms of each element present in a compound in terms of the smallest whole numbers.

The **actual weights** of substances involved in a chemical reaction are in the same ratio as their **equation weights.**

CHAPTER 6

GASES

Matter exists in three physical states: gaseous, liquid, and solid. A **gas** has no internal boundary. It expands to fill any container completely regardless of the size or shape of the container. A **liquid** has one internal boundary, its surface. It fills its container below its surface regardless of the shape of the container. A **solid** is rigid, that is, it bounds itself internally in all dimensions. It needs no external container.

The properties of these three states of matter are related to our concept of the structure of matter, and a study of them will help us become more familiar with the chemical behavior of matter. Let us first turn our attention to the gaseous state.

PRESSURE

The pressure of a gas is the force it exerts on a unit surface area. Pressure is measured with an instrument known as a **barometer.** This instrument consists of a glass tube about one meter long sealed at one end. It is filled completely with liquid mercury. The open end is then immersed in a dish of mercury by placing a thumb over the open end, inverting the tube, dipping the open end of the tube under the surface of mercury in the dish, and removing the thumb. The mercury then flows down in the tube until the pressure of the column of mercury in the tube is exactly equal to the pressure of air on the surface of the mercury in the dish. A void known as a **vacuum** is left in the sealed end of the tube.

Since the height of the mercury column depends upon the pressure of the air, the measurement of this height (see Figure 17) gives an indication of the air pressure. At sea level, the average air pressure supports a column of mercury 760 mm., or 29.92 in., high. This pressure is called **standard pressure.**

One mm. of mercury pressure is also called a **torr,** in honor of Evangelista Torricelli, who invented the first barometer.

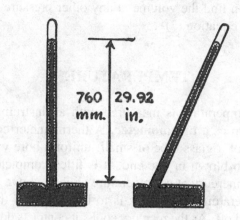

760 mm. 29.92 in.

Fig. 17. Mercury Barometers

If, at a given temperature, a gas is compressed, the volume of the gas will decrease. The English scientist Boyle studied this phenomenon carefully and found that **at a given temperature, the volume occupied by a gas is inversely proportional to the pressure.** This is known as **Boyle's Law.** Stated as an equation this would be:

$$P = k \frac{1}{V}. \qquad (1)$$

where:

P = pressure of a gas sample.
V = volume of a gas sample.
k = a constant.

Therefore:

$$P V = k \qquad (2)$$

So we see that, at a given temperature, the product of the pressure and volume of a gas must be constant. If the pressure is increased, the volume must decrease to maintain the constant product. For a given gas sample to be studied under different pressures, the following expression must hold:

$$P_1 V_1 = P_2 V_2 \qquad (3)$$

where:

P_1 = original pressure of a gas sample.

V_1 = original volume of the sample.
P_2 = new pressure of the sample.
V_2 = new volume of the sample.

If we know the volume of a gas at one pressure, we can find the volume at any other pressure by using equation (3).

TEMPERATURE

Temperature is measured with an instrument known as a **thermometer.** A thermometer consists of a glass tube of small, uniform bore with a bulb blown in one end. It is filled completely with mercury at a temperature slightly above the temperature at which it is to be used, and then sealed off. As the mercury cools, it contracts down into the tube, leaving a vacuum in the upper part of the tube. The thermometer must then be **calibrated** with known reference temperatures. The freezing point and boiling point of water are standard reference points. The thermometer is first immersed in melting ice, and the position of the end of the mercury column is marked with a line. The thermometer is then suspended in steam rising from water which is boiling at an atmospheric pressure of 760 mm., and the position of the end of the mercury column is again marked with a line. The distance between these two lines is then divided into equal lengths called **degrees.** The size of each degree depends upon the scale of temperature to be used.

There are two common temperature scales, Celsius, or Centigrade, and Fahrenheit. In the **Celsius scale,** denoted as ° C., the freezing point of water is called 0° C., and the boiling point of water is called 100° C. Thus this scale has 100 degrees between the two standard reference points. The Celsius scale is used in all scientific work.

In the **Fahrenheit scale,** denoted ° F., the freezing point of water is called 32° F., and the boiling point of water is called 212° F. Thus, in this scale there are 180 degrees between the two standard reference points. The Fahrenheit scale is used in the home and in industry. The two temperature scales are related by the following expression:

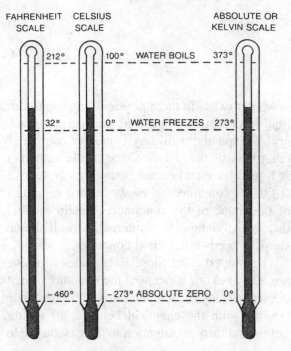

Fig. 18. Comparison of Temperature Scales

$$^\circ F. = \frac{9}{5} \,^\circ C. + 32. \qquad (4)$$

This expression can be used to convert a temperature from one scale to the other.

The French scientist Jacques Charles, in studying the relationship between the volume of a gas and its temperature, discovered that the volume of a gas increases by 1/273 for each degree Celsius its temperature is increased. From this he reasoned that a temperature of -273° C. is the lowest possible attainable temperature. He called this temperature **absolute zero,** and established the absolute or **Kelvin temperature scale,** which is related to the Celsius scale as follows:

$$K = ^\circ C. + 273. \qquad (5)$$

This expression is used in finding the absolute temperature when the Celsius temperature is known. Note that although there is an absolute zero to temperature, there is no known upper limit to temperature.

Charles' studies led to the discovery that **at a given pressure, the volume occupied by a gas is directly proportional to the absolute temperature of the gas.** This is known as **Charles'**

Law. Expressed as an equation this is:

$$V = kT \qquad (6)$$

where:

V = volume of the gas sample.
T = absolute temperature of the gas sample.
k = a constant.

Solving this expression for k, we find that the ratio of the volume of a gas to its absolute temperature is a constant:

$$\frac{V}{T} = k. \qquad (7)$$

Thus for a given gas sample, if the temperature is changed, this ratio must remain constant, so the volume must change in order to maintain the constant ratio. The ratio at a new temperature must be the same as the ratio at the original temperature, so:

$$\frac{V_1}{T_1} = \frac{V_2}{T_2} \qquad (8)$$

where:

V_1 = original volume of sample of gas.
T_1 = original absolute temperature.
V_2 = new volume of the sample.
T_2 = new absolute temperature of the sample.

If we know the volume of a gas at one absolute temperature, we can find its volume at any other temperature by using equation (8).

COMBINED GAS LAW

Boyle's Law and Charles' Law may be combined into one expression as follows:

$$\frac{P_1 V_1}{T_1} = \frac{P_2 V_2}{T_2} \qquad (9)$$

This expression is known as the **Combined Gas Law.** The temperature and pressure at which the volume of a gas is measured are known as the **conditions** of measurement. When the volume of a gas is known at one set of conditions, its volume at a new set of conditions can be found with the Combined Gas Law. Solving equation

(9) for the new volume, we have:

$$V_2 = V_1 \times \frac{P_1}{P_2} \times \frac{T_2}{T_1} \qquad (10)$$

This tells us that the new volume of the gas equals the old volume multiplied by two correction factors, one for pressure and the other for temperature. Note the following very closely. If the new pressure is greater than the original pressure, the effect on the volume will be to make it smaller. Therefore the pressure correction must be a fraction smaller than 1. Similarly, a new pressure less than the original will cause the gas to expand, thus making the correction greater than 1. With temperature, it works the other way. If the new temperature is higher than the original, the effect is to expand the gas. Thus the temperature correction must be greater than 1. If the new temperature is less than the original, the gas will contract, and the correction must then be less than 1. Always think through problems of this type before substituting into equation (10). This will help you avoid careless mistakes. Let us go through an example.

EXAMPLE 1: A gas sample occupies 250 cc. at 27° C. and 780 torr pressure. Find its volume at 0° C. and 760 torr pressure.

SOLUTION: First the temperatures must be converted to the kelvin scale by adding 273° to each:

$$T_1 = 27 + 273 = 300 \text{ K}$$
$$T_2 = 0 + 273 = 273 \text{ K}$$

Now, in the problem:
Temperature decreases, therefore volume decreases, therefore temperature correction must be less than 1.

Similarly,
Pressure decreases, therefore volume increases, therefore pressure correction must be greater than 1.

Therefore: $V_2 = 250 \times \dfrac{273}{300} \times \dfrac{780}{760}$.

Check this with equation (10)!
Solving:

$$V_2 = 234 \text{ cc.}$$

EXPERIMENT 8: Hold your thumb on the end of a small

hand bicycle pump, and pump the handle vigorously a few times with as much pressure as your thumb can hold. Feel the lower portion of the pump. Do you feel the increase in temperature? Repeat the pumping process, and this time feel the upper part of the pump. Do you note any cooling? The ratio of P to T for any gas must remain constant. At the lower end, where the pressure is increased, the temperature must also rise, and at the upper end, where the pressure is lowered, the temperature also drops.

STANDARD CONDITIONS

It can be seen from the previous discussion that a statement of the volume of a gas without specifying the conditions under which it is measured is meaningless. All gases fill their containers completely. For example, if we have a mixture of oxygen, nitrogen, and carbon dioxide gases and put a sample of this mixture in a one-cubic-foot container, how much, by volume, of each gas will be present? The answer is, one cubic foot of each, for each will fill the container. Thus we have a situation where:

1 cu. ft. of O_2 + 1 cu. ft. of N_2 + 1 cu. ft. of CO_2 = 1 cu. ft. of mixture.

This sounds absurd, but it is absurd only because we are talking about volumes of gases without specifying the conditions.

In order to compare gas volumes, a set of standard conditions has been established. The standard temperature is the freezing point of water. The standard pressure is the average pressure at sea level. Table XIII summarizes standard conditions in various units.

Table XIII	
Standard Conditions	
Temperature	*Pressure*
0° C.	760 torr
32° F.	29.92 in.
273 K	1.000 atmospheres

LAW OF PARTIAL PRESSURES

Dalton's Law of Partial Pressures states that **the total pressure of a mixture of two or more gases which do not chemically combine is the sum of the partial pressures of each.** In a mixture of gases, each gas exerts the same pressure as if it occupied the volume alone. Since, in a given gas mixture, the temperature of the gases is the same and all fill the same container, the relative amount of each gas present is indicated by its partial pressure. For example, it is usually stated that air is composed of approximately 21% oxygen by volume and 79% nitrogen by volume. This actually means that the partial pressure of oxygen in air is 21% of the total air pressure, and the partial pressure of nitrogen in air is 79% of the total air pressure. Thus we see that the volumetric composition of a gas mixture is indicated by the partial pressures of the constituents.

Insoluble gases are frequently collected by bubbling them into a container full of water to displace the water with the gas. In this process the gas sample becomes saturated with water vapor. The total pressure of the collected gas is the sum of the pressure of the dry gas and the pressure of the water vapor. The amount of dry gas collected can be found only after the vapor pressure of water at the temperature of collection is subtracted from the total gas pressure. Thus, Dalton's Law of Partial Pressures enables us, in a sense, to dry a gas by arithmetic.

EXPERIMENT 9: Select a wad of steel wool large enough to stick in the bottom of a jar when wet. Thoroughly moisten the steel wool in vinegar, press it into the bottom of the jar, and then invert the jar into a basin of water so as to entrap the air in the jar. If possible, clamp the jar in place. You will observe the rusting taking place in the steel wool, and the rising of the water in the inverted jar. In rusting, the steel removes oxygen from the air, thus reducing the pressure of the entrapped air, and permitting water from the basin to replace the oxygen. If the experiment is permitted to stand overnight, you will observe that the effect occurs only to a definite extent. When about one-fifth of the air has been replaced by water, the action will cease. This proves that air is made up of more than one gaseous substance, and that the total pressure is the sum of the partial pressures of each.

Table XIV

Vapor Pressure of Water

Temp. in °C	Press. in torr	Temp. in °C	Press. in torr	Temp. in °C	Press. in torr
0	4.6	22	19.8	40	55.3
5	6.5	23	21.1	45	71.9
10	9.2	24	22.4	50	92.5
15	12.8	25	23.8	60	149.4
16	13.6	26	25.2	70	233.7
17	14.5	27	26.7	80	355.1
18	15.5	28	28.4	90	525.8
19	16.5	29	30.0	100	760.0
20	17.5	30	31.8	200	11659.2
21	18.6	35	42.2	300	64432.8

Table XIV gives the vapor pressure of water in torr at various temperatures.

Now let us look at an example incorporating all the relationships studied thus far in this chapter.

EXAMPLE 2: 500 cc. of a gas is collected by the displacement of water at a temperature of 77° F. and a pressure of 748.8 torr. Find the volume of the dry gas at standard conditions.

SOLUTION: To use Table XIV we need the Celsius temperature of the gas. From Equation 4, p. 40:

$$77 = \frac{9}{5} C + 32$$

$$C = \frac{5}{9} (77 - 32) = 25° C.$$

From Table XIV we find that the vapor pressure of water at 25° C. is 23.8 torr. Therefore the pressure of dry gas is:

$$748.8 - 23.8 = 725 \text{ torr}$$

Standard conditions are 0° C. and 760 torr pressure. Converting our temperatures to the absolute scale we have:

$$T_1 = 25 + 273 = 298 \text{ K}$$
$$T_2 = 0 + 273 = 273 \text{ K}$$

Temperature decreases, therefore correction factor is less than 1.
Pressure increases, therefore correction factor is less than 1. Thus:

$$V_2 = 500 \times \frac{273}{298} \times \frac{725}{760} = 437 \text{ cc. of dry gas.}$$

Problem Set No. 6

1. Convert 30° C. to ° F.
2. Convert 68° F. to ° C.
3. Convert 95° F. to K.
4. 500 cc. of gas at 770 torr pressure is compressed to 1540 torr pressure, at constant temperature. Find the new volume of the gas.
5. 350 cc. of gas at 47° C. is heated to 188° C. at constant pressure. Find the new volume of the gas.
6. A sample of gas at 20° C. and 750 torr is heated to 252° C. while the volume is held constant. Find the new pressure of the gas.
7. The volumetric analysis of a sample of gas leaving a smoke stack is as follows:

    10.8% CO_2 2.2% CO 4.5% O_2 82.5% N_2

    The pressure of the gas sample is 750 torr. Find the partial pressure of each of the constituents.
8. 150 cc. of a gas is collected at 22° C. and 740 torr. Find the volume of the gas at standard conditions.
9. A gas occupies 330 cc. at standard conditions. Find its volume at 86° F. and 30.40 in. pressure.
10. 400 cc. of gas is collected over water at 24° C. and 767.4 torr. Find its volume at standard conditions when dry.

KINETIC MOLECULAR THEORY

Some of the principal general properties of gases may be listed as follows:

1. Gases are compressible.
2. Gases fill any container.
3. Different gases mix completely.
4. Gases expand on heating.
5. Gases do not settle in their container.

The Kinetic Molecular Theory was developed as scientists attempted to find an explanation of these properties. According to this theory, gases consist of tiny, discrete molecules. In gases, these molecules are relatively far apart with empty space between them. This explains the ease with which gases can be compressed. The gas molecules are in constant, rapid motion. They move in straight lines until they collide with other molecules, or with the walls of the container. This explains the filling of containers by gases and the mixing of gases. The moving gas molecules, in colliding with the walls of the container cause the pressure on the container. A given pressure is the result of the number of such collisions in a unit time. Gas pressure is increased by:

    a. Forcing more gas into the container—thus increasing the number of collisions per unit time.

    b. Decreasing the volume of the gas—thus shortening the average distance between the molecules and thereby increasing the number of collisions per unit time.

    c. Heating the gas in a closed container—thus increasing the speed of the molecules and thereby increasing the number of collisions per unit time.

The speed of the moving molecules is the result of the **kinetic energy** (energy of motion) they possess. This kinetic energy is increased by heating the gas and decreased by cooling it. The Kinetic Molecular Theory suggests that the collisions of the gas molecules with other molecules or with the walls of the container are perfectly **elastic**—that is, they take place without loss of energy either through friction or through any other means. If there were energy losses as a result of these collisions, a loss in kinetic energy would result, and the gas would ultimately settle in its container. The concept of elastic collisions explains the fact that gases do not settle.

IDEAL GASES

An **ideal gas** is one which follows the gas laws perfectly. Such a gas is nonexistent, for no known gas obeys the gas laws at all possible temperatures. Consider for a moment what would happen if we were to cool a gas to absolute zero. The gas would vanish!

$$V_2 = V_1 \times \frac{P_1}{P_2} \times \frac{0}{T_1} = 0.$$

This, of course, does not happen in nature. There are two principal reasons why real gases do not behave as ideal gases:

    1. The molecules of a real gas have mass, or weight, and the matter thus contained in them cannot be destroyed.

    2. The molecules of a real gas occupy space,

and thus can be compressed only so far. Once the limit of compression has been reached, neither increased pressure nor cooling can further reduce the volume of the gas.

In other words, a gas would behave as an ideal gas only if its molecules were true mathematical points—that is, if they possessed neither mass nor dimensions. However, at the ordinary temperatures and pressures used in industry or in the laboratory, molecules of real gases are so small, have so little mass, and are so widely separated by empty space, they behave almost as if they were mathematical points—that is, they follow the gas laws so closely that any deviations from these laws are insignificant. Nevertheless, it should be borne in mind that the gas laws are not strictly accurate, and results obtained from them are really close approximations.

DENSITY OF GASES

The **absolute density** of a gas is the mass in grams of one milliliter of the gas **at standard conditions.** Compared with liquids and solids, gases have very low density. This is the result of the relatively large amount of empty space between gas molecules as suggested by the Kinetic Molecular Theory. Hydrogen has a density of 0.00009 g./ml. Oxygen has a density of 0.001429 g./ml. These figures are found experimentally by finding the actual mass of a known volume of a gas sample at known temperature and pressure conditions. The density is then computed with the aid of the Combined Gas Law. Let us look at an example.

EXAMPLE 3: 564.3 ml. of chlorine gas at 27° C. and 740 torr has a mass of 1.607 g. Find the density of chlorine.

SOLUTION: First the volume must be reduced to standard conditions.

$$V_2 = 564.3 \times \frac{273}{300} \times \frac{740}{760} = 500 \text{ ml.}$$

The mass of chlorine, of course, is not changed, for we are still dealing with the same amount of it. Density is the mass of one ml. of the gas at standard conditions. Since 500 ml. of chlorine at standard condi-

Table XV

Density of Gases

Gas	Formula	Absolute Density in g./ml.	Relative Density Air = 1
Acetylene	C_2H_2	0.001173	0.9073
Air	——	0.0012929	1.0000
Carbon dioxide	CO_2	0.0019769	1.5290
Carbon monoxide	CO	0.0012504	0.9671
Chlorine	Cl_2	0.003214	2.486
Helium	He	0.00017847	0.13804
Hydrogen	H_2	0.00008988	0.06952
Nitrogen	N_2	0.00125055	0.96724
Oxygen	O_2	0.001429	1.10527
Sulfur dioxide	SO_2	0.002927	2.2639

tions has a mass of 1.607 g., we can find the mass of 1 ml. with a simple proportion thus.

$$\frac{500}{1.607} = \frac{1}{x}$$

$$x = 0.003214 \text{ g.}$$

Therefore, the density of chlorine is 0.003214 g./ml.

Since the absolute densities of gases involve such small numbers, they are commonly compared with the density of air. Such a comparison, indicating the number of times a gas is more dense than air, is called the **relative density of the gas.** Table XV gives the absolute and relative densities of several common gases.

LAW OF DIFFUSION

Gas molecules are small enough to diffuse through such porous materials as unglazed porcelain and natural rubber. The English scientist Graham discovered that **the rates of diffusion of gases are inversely proportional to the square roots of their densities,** when the gases are at the same temperature and pressure. This is known as Graham's Law of Diffusion. Expressed as an equation this would be:

$$\frac{rate_1}{rate_2} = \sqrt{\frac{density_2}{density_1}}$$

This law is particularly useful in comparing rates of diffusion of different gases. Let us look at an example.

EXAMPLE 4: How much faster will hydrogen diffuse than helium?

SOLUTION: From Table XV we find the densities:

Helium   0.00018 g./ml.
Hydrogen   0.00009 g./ml.

Therefore:

$$\frac{\text{Hydrogen rate}}{\text{Helium rate}} = \sqrt{\frac{0.00018}{0.00009}} = \sqrt{2} = 1.4$$

Thus, hydrogen diffuses 1.4 times as fast as helium.

This law was put to use in the early work of preparing the original atomic bomb. Uranium metal was combined with fluorine to form the gas UF_6. Molecules of this gas containing the isotope of uranium of atomic weight 235 (U-235) diffused slightly more rapidly through a porous membrane than molecules of the gas containing uranium isotopes of atomic weight 238 (U-238). By repeated diffusions, U-235, the bomb material, was separated from the useless U-238.

GRAM MOLECULAR VOLUME

From the density of a given gas, we can find out what volume a gram-mole of the gas would occupy at standard conditions through the use of a simple proportion. For example, in the case of hydrogen this would be:

Density of hydrogen   0.00009 g./ml.
Gram molecular weight   2.016 g.

So:

$$\frac{0.00009}{1} = \frac{2.016}{x}$$

$$x = 22,400 \text{ ml.} = 22.4 \text{ liters.}$$

Similar calculations for the other gases yield approximately the same results. Thus, we may conclude that **for any gas, the volume occupied at standard conditions by one gram-mole of the gas is 22.4 liters.**

This is a significant fact, for now we can see that when we speak about a mole of a gas, we are referring to both a definite weight of a gas (its molecular weight) and a definite volume of it at standard conditions (22.4 liters). This dual meaning of the mole is one of the most important concepts in chemistry.

AVOGADRO'S NUMBER

We have seen that Avogadro's Hypothesis states that equal volumes of gases contain equal numbers of molecules. Since, at standard conditions, one gram-mole of any gas occupies the same gram molecular volume, one gram-mole of every molecular substance must contain the same number of molecules. This number has been found by a variety of experimental methods. It is **6.022 × 10²³**. As you can see, this is a very large number. It is called Avogadro's Number. From it we can find the actual weight of a molecule of a substance. For example:

6.022×10^{23} molecules of O_2 weigh 32 g.

Therefore: 1 molecule of O_2 weighs:

$$\frac{32}{6.022 \times 10^{23}} = 5.31 \times 10^{-23} \text{ g.}$$

This is:

  0.000 000 000 000 000 000 000 053 1 grams.

Since the molecule of oxygen contains 2 atoms, each atom of oxygen must weigh half this amount, or 2.65×10^{-23} grams. Amazing as it may seem, numbers like this are very important in science.

FINDING MOLECULAR WEIGHTS

To find the molecular weight of a substance, we simply weigh a known volume of it in the gaseous state at known temperature and pressure, reduce the volume to standard conditions, and then calculate the weight of 22.4 liters at standard conditions using a simple proportion. The direct weighing of gases is often quite difficult. Therefore, in actual practice, a solid or liquid substance which will decompose to produce the desired gas is weighed both before and after decomposition. The loss in weight is then the weight of the gas. Oxygen can be formed by heating solid compounds such as potassium chlorate, $KClO_3$. Carbon dioxide can be produced by heating calcium carbonate, $CaCO_3$. Sulfur dioxide can be formed by treating solid sodium sulfite, Na_2SO_3, with an acid.

Let us look at typical experimental results obtained in a laboratory in attempting to find the molecular weight of oxygen by heating $KClO_3$.

EXAMPLE 5: 7.00 g. of $KClO_3$ are heated until 1.55 liters of oxygen are produced at 27° C. and 756 torr. The $KClO_3$ residue then weighs 5.00 g. Find the molecular weight of oxygen.

SOLUTION:
Weight of oxygen: 7.00 − 5.00 = 2.00 g.
Volume of oxygen at standard conditions:

$$V_2 = 1.55 \times \frac{273}{300} \times \frac{756}{760} = 1.40 \text{ liters.}$$

Therefore, by proportion, the molecular weight of oxygen is:

$$\frac{1.40}{2.00} = \frac{22.4}{x}$$

$$x = \frac{22.4 \times 2.00}{1.40} = 32.0 \text{ g.}$$

TRUE FORMULAS

We saw in the last chapter that we could compute the simplest formula of a substance if we knew its percentage composition. However, we were unable to find the true formula of molecular compounds. The **true formula** gives the exact number of atoms of each element in a molecule of the substance. For molecular substances, the true formula must show two things:

1. The ratio of the number of atoms of each element present.
2. The molecular weight of the substance (the sum of the atomic weights in the formula).

Therefore in order to find the true formula of a substance we must know both its percentage composition and its molecular weight. Let us look at an example to see how true formulas may be found.

EXAMPLE 6: 500 ml. of a gaseous substance at standard conditions weigh 0.58 g. The substance contains 92.31% carbon and 7.69% hydrogen. Find its true formula.

SOLUTION: First we find the molecular weight, that is, the weight of 22.4 liters of the substance at standard conditions.

$$\frac{500}{0.58} = \frac{22400}{x}$$

$$x = \frac{22400 \times 0.58}{500} = 26 \text{ g.}$$

The molecular weight of the substance is thus 26. Next we find the simplest formula:

Hydrogen: $\frac{7.69}{1} = 7.69$; $\frac{7.69}{7.69} = 1$.

Carbon: $\frac{92.31}{12} = 7.69$; $\frac{7.69}{7.69} = 1$.

Thus the simplest formula is C_1H_1, or simply CH. The weight represented by this simplest formula is $12 + 1 = 13$.

If we divide the true molecular weight by the formula weight of the simplest formula we get the number **which must multiply each of the subscripts in the simplest formula** to give the true formula. Thus:

$$\frac{26}{13} = 2.$$

Therefore the true formula is: C_2H_2.
Checking its weight we see that the formula weight is now $24 + 2 = 26$, which equals the molecular weight of the substance.

MASS–VOLUME RELATIONSHIPS

Since, for gases, the mole represents both a mass and a volume, we can apply this concept to equations to find out the volumes at standard conditions of gaseous reactants or products. Let us look at an example.

EXAMPLE 7: Oxygen is liberated from potassium chlorate by heat according to the equation:

$$2 KClO_3 = 2 KCl + 3 O_2$$

What volume of O_2 at standard conditions can be produced by the complete decomposition of 6.13 g. of $KClO_3$?

SOLUTION: From the equation we see that 2 moles of $KClO_3$ form 3 moles of O_2, or, in other words, for every mole of $KClO_3$ decomposed 3/2 moles of O_2 are formed. A mole of $KClO_3$ contains 122.6 g. (sum of the atomic weights). Therefore, we are going to decompose 6.13/122.6 moles of $KClO_3$. The moles of O_2 thereby produced will be:

$$\frac{6.13}{122.6} \times \frac{2}{3} = 0.075 \text{ moles of } O_2.$$

But at standard conditions each mole of O_2 occupies 22.4 liters. Therefore the volume of O_2 produced at standard conditions must be:

$$22.4 \times 0.075 = 1.68 \text{ liters of } O_2.$$

To find this volume at other than standard conditions, we can then make use of the gas laws. Note that all volumes found from equations apply only to standard conditions!

VOLUME–VOLUME RELATIONSHIPS

Since, at standard conditions, a mole represents a definite volume of a gas, a ratio of moles of gases must indicate the same ratio of volumes of gases. Consider the following equation in which all the substances are gases:

$$N_2 \quad + \quad 3\,H_2 \quad = \quad 2\,NH_3$$

Nitrogen		Hydrogen		Ammonia
1 mole	:	3 moles	:	2 moles
1 volume	:	3 volumes	:	2 volumes

Note that the molar ratio is identical to the volumetric ratio of the substances so long as all the volumes are measured **at the same conditions.** Let us look at an example involving the volumes of gases in an equation.

EXAMPLE 8: 60 liters of hydrogen, measured at room temperature and pressure, are to be used in the preparation of ammonia gas.

   (a) What volume of nitrogen, measured at the same conditions, will be required for the reaction?

   (b) What volume of ammonia, measured at the same conditions, can be prepared?

SOLUTION: The equation is:

$$N_2 \quad + \quad 3\,H_2 \quad = \quad 2\,NH_3$$

$$1 \text{ vol.} \qquad 3 \text{ vol.} \qquad 2 \text{ vol.}$$

Part a: The volume of nitrogen required is:

$$\frac{x}{1} = \frac{60}{3}$$

$$x = 20 \text{ liters of nitrogen.}$$

Part b: The volume of NH_3 which can be produced is:

$$\frac{x}{2} = \frac{60}{3}$$

$$x = 40 \text{ liters of ammonia.}$$

EQUIVALENT WEIGHT OF AN ELEMENT

The equivalent weight of an element may be defined as that amount of it which has combined with or displaced one atomic weight's worth (1.0 grams) of hydrogen. It is also the amount of an element which has combined with or displaced 8 parts by weight (8 grams) of oxygen. A quantity of an element equal in amount to its equivalent weight is called **an equivalent** of the element.

Just as we have been referring to a mole of a substance, we are now ready to refer to an equivalent of a substance. These two ideas—for they are ideas rather than mere defined quantities—form the basis of all quantitative applications of chemistry. We have seen how the idea of the mole is used with chemical equations. The coefficients of a balanced equation give the ratio of the number of moles of reactants and products involved in the reaction. Equivalents likewise refer to amounts of reactants and products. The rule is simply this. **One equivalent of any substance in nature reacts actually or theoretically with one equivalent of every other substance to produce one equivalent of each of the products involved.**

The concept of the equivalent is thus a powerful tool for use in prying open the secrets of chemistry. All substances combine, or are produced, on an "equivalent" basis. But just like any other tool, this one must be used skillfully and with understanding of its application. Let us focus our attention now on how to find the equivalent weight of an element.

The definitions of equivalent weight suggest experimental methods of measuring the equivalent weight of an element. The mass of an element which combines with or displaces either 1 gram of hydrogen or 8 grams of oxygen would be the equivalent weight of the element in grams. Let us look at some examples.

EXAMPLE 9: Many metals displace hydrogen from acids. At standard conditions, one mole of hydrogen occupies 22.4 liters. But a mole of hydrogen contains 2 atomic weights' worth of hydrogen, for the formula of hydrogen is H_2. Therefore 11.2 liters of hydrogen at standard conditions would be 1 atomic weight's worth of it. Thus, that mass of a metal which would displace 11.2 liters of hydrogen at standard conditions would be the equivalent weight of the metal.

   1.50 g. of a metal (zinc) displace 560 ml. of hydrogen at 20° C. and 748 torr pressure. Find the equivalent weight of the metal.

SOLUTION: First we find the volume of hydrogen at standard conditions.

$$V_2 = 560 \times \frac{273}{293} \times \frac{748}{760} = 513 \text{ ml.}$$

Then, by proportion:

$$\frac{1.5}{513} = \frac{x}{11200}$$

$$x = 32.7 \text{ g. of the metal.}$$

Thus, the equivalent weight of the metal is 32.7.

EXAMPLE 10: Oxygen combines directly with most elements to form oxides. The analysis of an oxide to find its percentage composition gives a ratio of weights of oxygen and the other element present. In fact, the percentage composition is the number of grams of each element present in a 100 gram sample. Therefore, knowing the percentage composition we can use a simple proportion to find the weight of an element which has combined with 8 grams of oxygen.

Zinc oxide contains 80.34% zinc and 19.66% oxygen. Find the equivalent weight of zinc.

SOLUTION: By proportion:

$$\frac{80.34}{19.66} = \frac{x}{8}$$

$$x = 32.7 \text{ g}.$$

Thus the equivalent weight of zinc is 32.7.

With the techniques suggested by the foregoing examples, we can find the equivalent weight of many elements. Table XVI gives both the atomic weight and the equivalent weight of some selected elements together with the ratio of the atomic weight to the equivalent weight in each case.

Table XVI
Equivalent Weights

Element	Atomic Wt.	Equivalent Wt.	Atomic Wt. Equiv. Wt.
Hydrogen	1.008	1.008	1
Sodium	22.990	22.990	1
Potassium	39.098	39.098	1
Silver	107.87	107.87	1
Oxygen	15.999	8.000	2
Magnesium	24.31	12.16	2
Calcium	40.08	20.04	2
Zinc	65.38	32.69	2
Aluminum	26.98	8.99	3

Look carefully at the ratio in the last column for each element. What is it? Yes, it is the **valence number** of the element. This is very important. We may set down the following rule:

$$\text{Valence Number} = \frac{\text{Atomic Weight}}{\text{Equivalent Weight}}$$

or:

$$\text{Equivalent Weight} = \frac{\text{Atomic Weight}}{\text{Valence Number}}$$

You will recall that an element exhibits a valence number other than zero only when it is in a compound. Furthermore, many elements are capable of exhibiting more than one valence number. The particular valence number exhibited by an element depends upon the compound under consideration. Since equivalent weight is related to valence number, these same points apply to it. Elements which can have different valence numbers possess equivalent weights corresponding to each of its valence numbers. The equivalent weight of an element depends upon the valence number to be shown by the element in a particular reaction.

Let us look at the reactions of iron with oxygen. Depending upon the conditions under which the reaction is to take place, iron may form two different oxides according to the following equations:

$$2 \text{ Fe} + \text{ O}_2 = 2 \text{ FeO}.$$

$$4 \text{ Fe} + 3 \text{ O}_2 = 2 \text{ Fe}_2\text{O}_3.$$

In both of these oxides, oxygen has a valence number of -2. In FeO, the valence number of iron is $+2$. In Fe_2O_3, the valence number of iron is $+3$. Iron is exhibiting two different valence numbers, depending upon the particular reaction under consideration. Therefore iron has two different equivalent weights. In the first reaction, the equivalent weight of iron is one-half its atomic weight. In the second reaction, the equivalent weight of iron is one-third its atomic weight. You can thus see that, for elements which exhibit more than one valence number, the equivalent weight of the element depends upon which of the valence numbers the element will use in the particular reaction involved. The elements in Table XVI were selected because they normally have but one valence number, and consequently have but one equivalent weight.

RULE OF DULONG AND PETIT

The quantity of heat absorbed by a substance in warming up, or given off by it in cooling, is specified in units known as calories. A **calorie** is that quantity of heat necessary to raise the temperature of one gram of water one degree Celsius. The number of calories of heat necessary to raise the temperature of one gram of any substance one degree Celsius is known as the **heat capacity** of the substance. The ratio of the heat capacity of a given substance to the heat capacity of water is known as the **specific heat** of the substance.

The French scientists Dulong and Petit studied the relationship between specific heat and atomic weight of elements. They discovered the following relationship for elements in the solid state:

$$\text{Atomic Weight} \times \text{Specific Heat} = \text{Approximately } 6.2$$

This product is known as the **atomic heat** of the element. The figure 6.2 is an average figure. Table XVII gives the atomic heat of several elements.

Table XVII Atomic Heat			
Element	*Atomic Wt.*	*Specific Heat*	*Atomic Heat*
Lithium	7	0.850	6.0
Sodium	23	0.292	6.7
Magnesium	24	0.245	5.9
Phosphorus	31	0.181	5.6
Potassium	39	0.180	7.0
Calcium	40	0.155	6.2
Iron	56	0.108	6.0
Copper	64	0.092	5.9
Zinc	65	0.092	6.0
Silver	108	0.056	6.0
Tin	119	0.054	6.4
Gold	197	0.031	6.1

The exact atomic weight of an element may be found by using the Rule of Dulong and Petit in conjunction with methods of finding the equivalent weight of an element. To do this we proceed in four steps:

1. Determine the exact equivalent weight.
2. Find the approximate atomic weight from the Rule of Dulong and Petit.
3. Find the valence number by dividing the approximate atomic weight by the exact equivalent weight.
4. Multiply the equivalent weight by the valence number to obtain the exact atomic weight.

Let us look at an example.

EXAMPLE 11: A compound contains 79.87% copper and 20.13% oxygen. The specific heat of copper is 0.093. Find the exact atomic weight of copper.

SOLUTION:

Step 1. The exact equivalent weight of copper (that weight of it combined with 8 grams of oxygen) is:

$$\frac{79.87}{20.13} = \frac{x}{8}$$

$$x = 31.74 \text{ g.}$$

Thus the equivalent weight of copper is 31.74.

Step 2. From the Rule of Dulong and Petit:

$$\text{Appr. Atomic Wt. of Copper} = \frac{6.2}{\text{Sp. Heat}} = \frac{6.2}{0.092}$$
$$= 67.4.$$

Step 3. The valence number must be a whole number. Dividing the approximate atomic weight by the equivalent weight we have:

$$\frac{67.4}{31.74} = 2.12$$

The valence number is the nearest whole number, which is 2.

Step 4. Multiplying the equivalent weight by the valence number we have:

$$31.74 \times 2 = 63.5$$

The exact atomic weight of copper is therefore 63.5.

Problem Set No. 7

1. 555 ml. of a gas at 22° C. and 740 torr pressure weigh 0.6465 g. Find the density of the gas.
2. How much faster will helium diffuse than chlorine?

3. Find the actual weight in grams of a single hydrogen atom.

4. 0.24 g. of a gas occupy 82.2 ml. at 117° C. and 740 torr pressure. Compute its molecular weight.

5. A gas consists of 85.72% carbon and 14.28% hydrogen. 0.855 g. of this gas occupies 523 ml. at 29° C. and 733 torr pressure. Find the true formula of the substance.

6. What volume of hydrogen, at standard conditions, can be produced by treating 10 g. of zinc with hydrochloric acid, HCl?

7. What weight of calcium carbonate, $CaCO_3$, must be heated in order to produce 5 liters of CO_2 at standard conditions?

8. What volume of pure O_2 will be required to burn 15 liters of ethane, C_2H_6, to CO_2 and H_2O? (all gases are measured at standard conditions).

9. Compute the equivalent weight of phosphorus in the compound phosphine, PH_3.

10. The oxide of an unknown element contains 10.2% oxygen and 89.8% of the element. The specific heat of the element is 0.0305. Find its exact atomic weight.

SUMMARY

Boyle's Law: At a given temperature, the volume occupied by a gas is inversely proportional to the pressure.

Charles' Law: At a given pressure, the volume occupied by a gas is directly proportional to the absolute temperature.

Combined Gas Law:

$$\frac{P_1 V_1}{T_1} = \frac{P_2 V_2}{T_2}$$

Standard conditions: The standard temperature is 0° C. The standard pressure is 760 torr.

Law of Partial Pressures: The total pressure of a mixture of two or more gases which do not chemically combine is the sum of the partial pressures of each.

The **Kinetic Molecular Theory** holds that gases consist of discrete molecules relatively far apart in rapid motion which collide with one another and with the walls of the container with perfect elasticity.

Real gases obey the gas laws only approximately because their molecules have mass and occupy space.

The **absolute density** of a gas is the mass of one milliliter of it at standard conditions.

The **relative density** of a gas is the ratio of its absolute density to the absolute density of air.

Law of Diffusion: The rates of diffusion of gases are inversely proportional to the square roots of their densities.

Gram-molecular volume: The volume occupied at standard conditions by one gram-mole of any gas is approximately 22.4 liters. The number of molecules in a gram-mole is 6.022×10^{23}. This is known as **Avogadro's number.**

The **equivalent weight** of an element is that weight of it which combines with or displaces 1.0 gram of hydrogen or 8 grams of oxygen.

Rule of Dulong and Petit:

Atomic Weight $\times$ Specific Heat = Approximately 6.2

CHAPTER 7

LIQUIDS AND SOLIDS

If the volume of a gas is sufficiently reduced by compressing it or cooling it or both, the gas will condense to a liquid. Early scientific investigators discovered that a number of substances in the gaseous state at room temperature could be condensed to a liquid by pressure alone. Other gases resisted **liquefaction** regardless of the pressure imposed, and condensed only after the temperature had been reduced. This led to the idea that a critical temperature was involved in the liquefaction process. The **critical temperature** of a gaseous substance is the temperature above which it is impossible to liquefy the substance by pressure alone. The pressure required to liquefy a gas at its critical temperature is called the **critical pressure.** Table XVIII gives the critical temperature and pressure of several common substances.

Table XVIII
Critical Temperatures and Pressures

Substance	Critical Temp. in °C.	Critical Press. in Atmospheres
Ammonia	132.4	111.3
Argon	−122.4	48.0
Carbon dioxide	31.0	72.9
Carbon monoxide	−140.2	34.5
Chlorine	144.0	76.1
Ethyl alcohol	243.1	63.0
Helium	−268.0	2.26
Hydrogen	−239.9	12.8
Nitrogen	−146.9	33.5
Oxygen	−118.4	50.1
Sulfur dioxide	157.5	77.9
Water	374.2	218.3

The more a gas is cooled below its critical temperature, the less pressure will be required to liquefy it. When a substance is in the gaseous state at a temperature above its critical temperature, it is properly called a **gas.** On the other hand, a substance in the gaseous state at a temperature below its critical temperature is properly referred to as a **vapor.**

VAPOR PRESSURE

All liquids show a tendency to vaporize at room temperature and pressure. If a small sample of a liquid is placed in a dish and allowed to stand in contact with freely moving air, it will in time completely **evaporate,** which means that its molecules have escaped as a vapor. Liquids like gasoline and carbon tetrachloride evaporate much more rapidly than liquids like oil or mercury. The rate of evaporation is basically related to a property of liquids known as **vapor pressure.**

To understand what vapor pressure is, think of a milk bottle partially filled with water with the cap on. The water starts to evaporate into the air in the bottle. The pressure of this air thus increases as the water vapor is added to it. Ultimately the water vapor will **saturate** the air, and droplets of water will begin to **condense** back to a liquid on the upper walls of the bottle. **The increase in pressure of the air when it is saturated with water vapor is known as the vapor pressure of the water.** The vapor pressure of other liquids can be found in the same way. For accurate measurement, the air to be used in the container in contact with the liquid would initially have to be free of any molecules of the liquid vapor. Needless to say, the higher the vapor pressure, the more rapidly a liquid will evaporate into dry air.

Since it takes energy for a molecule of a liquid to escape as vapor from the liquid, we can see that the vapor pressure of a liquid will be related to the temperature. Table XIV on page 43 gives the vapor pressure of water at various temperatures. Note particularly the vapor pressure of water at 100° C. You will see that it is 760 torr. This is standard atmospheric pressure, and 100° C. is called the **normal boiling point** of water. We can thus see that the **boiling point of a liquid is the temperature at which the vapor pressure of the liquid equals the pressure of the atmosphere about it.** The normal boiling point of a liquid is the temperature at which its vapor pressure reaches standard atmospheric pressure, or 760 torr.

Since atmospheric pressure changes slightly from day to day, the boiling point of water will fluctuate correspondingly. As the air pressure goes down, the boiling point will decrease. In places of high elevation the boiling point of water would be considerably reduced because of low atmospheric pressure. For example, in Denver, Colorado, water would boil at about 87° C., while at the top of Pike's Peak water would boil at about 62° C. Table XIX gives the normal boiling points of some common substances.

Table XIX
Normal Boiling Points

Substance	Temp. °C.	Substance	Temp. °C.
Benzene	80.1	Hydrogen	−252.8
Carbon dioxide	−78.5	Oxygen	−183.0
Chlorine	−34.1	Sulfur dioxide	−10.0
Ethyl alcohol	78.5	Water	100.0

EVAPORATION

The phenomenon of evaporation is familiar to all, and yet a few words should be said about it. Like boiling, it is a case of a liquid changing to a vapor. Unlike boiling, it takes place at any temperature.

To understand evaporation we need to know a bit about energy. According to the Kinetic Molecular Theory, the molecules of a liquid are free to move about under the surface of the liquid. The motion is less rapid than in the case of gases because the molecules of the liquid are virtually "in contact" with one another. But the molecules do move, for they possess kinetic energy (energy of motion) at any given temperature.

The amount of this energy possessed by each molecule of the liquid is not uniform. Some molecules possess considerably more energy than others in the same sample of a liquid. In general, most of the molecules have about the same amount of energy, but some possess appreciably more than average energy and some possess appreciably less. Furthermore, because the molecules are so close to one another, they collide frequently, and during such collisions energy is redistributed between the colliding molecules causing a gain in energy in one of the molecules and a loss in energy in the other. This transfer of energy by collision can result in the formation of relatively high-energy molecules.

Now the temperature is a measure of the **average energy** of a sample of liquid. If the temperature goes up, the average energy of the liquid goes up because warming indicates an increase in energy. Similarly, a decrease in average energy results in a lowering of the temperature of the sample.

The escape of a liquid molecule in the evaporation process requires energy. Consequently only those molecules of high energy can evaporate from a liquid. But the escape of these high energy molecules results in a lowering of the average energy of the liquid sample. Thus **a cooling always accompanies evaporation.**

EXPERIMENT 10: With pieces of cotton, spread water on the back of one hand and alcohol on the back of the other. Then wave your hands back and forth through air. You will observe cooling from the evaporation of both liquids, but since the alcohol is more volatile, it will cause more cooling.

Of course, as soon as the temperature of the liquid begins to drop below the temperature of the surrounding atmosphere, the atmosphere begins to warm the sample. This means that the atmosphere begins to add more energy to the liquid to replace the energy lost by evaporation. This additional energy, plus the accumulation of it by relatively few of the liquid molecules through molecular collisions, permits the evaporation of the liquid to continue until ultimately all of the molecules of the liquid have gained enough energy to evaporate.

You have undoubtedly observed the cooling effect of liquids like water or alcohol evaporating from your skin. You perspire when you get hot, and the evaporating perspiration acts as a check to prevent your body from overheating. The temperature of the air is usually cooler by the seashore, or at the lake, or by a waterfall, because the evaporating water absorbs energy from the atmosphere and cools it. Until recently, the dairy industry was confined to the northern states or to places of high elevation because dairy cattle did not possess sweat glands and consequently could not survive in the high temperatures of the South. However, by crossbreeding standard dairy cattle with cattle possessing sweat glands, scientists have been able to produce new varieties of dairy cattle which possess sweat glands and can live in more tropical climates. This, of course, has bolstered the economy of warmer regions. These illustrations are given to reemphasize the fact that natural phenomena follow definite laws of nature. Human progress is enhanced if we first learn these laws and then devise means of applying them in the control of our environment.

SURFACE TENSION

We have seen that the molecules of a liquid are relatively close together. They are so close, in fact, that they exert appreciable attraction on one another. Inside the body of the liquid, where a given molecule is completely surrounded by other molecules, these attractive forces are equal in all directions. Thus they counteract one another, and no net unbalanced force remains on the molecule. However, on the surface of the liquid a molecule is not completely surrounded by nearby molecules. The region above such a surface molecule is relatively vacant. Such a molecule is attracted by its neighbors in the surface and by the molecules below it. This results in a net unbalanced attractive force directed into the interior of the liquid. The effect is that the surface molecules form an encasing film on the liquid which is relatively quite tough.

You can readily observe the effects of surface tension. A tiny droplet of water remains intact and does not spread out to form a smooth film on a table top because the surface film holds the water back from "seeking its own level." Drops of water falling free and passing through air are drawn up into spheres by the surface film. A sphere, of course, is a shape containing a maximum volume within a minimum surface. You can float razor blades or needles on water even though these objects are far more dense than water because the tough surface film supports them.

EXPERIMENT 11: Bend the end of a 4-inch length of wire into a single loop around a pencil, and then bend the loop so that its plane is perpendicular to the remaining length of wire. Fill a glass with water. Using the straight part of the wire as a handle, gently press the flat loop against the surface of the water. Note how the surface becomes indented before the loop finally breaks through into the water. This is the result of the tough film on the water's surface. Now slowly pull the loop up through the surface. Note how the water is pulled up above the level of the remaining surface.

The surface tension of a liquid may be changed by dissolving substances in the liquid. For example, soap greatly reduces the surface tension of water. This can be observed as follows:

EXPERIMENT 12: Sprinkle a little black pepper onto the surface of some clean water in a basin. The particles of pepper will be supported by the surface film. Now touch a piece of soap to the surface of the water. As the soap dissolves, you will see the particles of pepper rapidly drawn away from the point at which the soap has weakened the film.

The decrease in the horizontal forces of attraction in the original water surface caused by the dissolving soap result in the formation of net unbalanced forces in the surface directed away from the point of application of the soap. The receding film then carries the pepper with it.

CAPILLARY ACTION

Surface tension is related to another phenomenon associated with liquids. Between the molecules of a liquid there are attractive forces which may be called forces of **cohesion.** Similarly there are attractive forces between the molecules of a liquid and the molecules of its container. These may be called forces of **adhesion.** If the forces of adhesion are greater than the forces of cohesion, the liquid will *wet* the container. If the reverse is true, the liquid will draw away from the container and not wet it. Figure 19 illustrates the difference in appearance of the surface of water and mercury in glass containers. Water wets glass. Mercury does not.

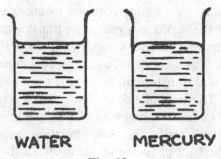

WATER   MERCURY

Fig. 19.
Surface Variation of Liquids
with Different Ability to Wet Glass

If a liquid wets a given substance, the liquid will be drawn up into a small-diameter tube made of the substance. If the liquid does not wet a substance, a small-diameter tube made of the substance will depress the surface of the liquid. Figure 20 illustrates this with water and mercury in glass tubes. The rise or depression of a liquid surface in a small-diameter tube is known as **capillary action.** The amount of change in the level of the liquid in the tube is directly proportional to the surface tension of the liquid.

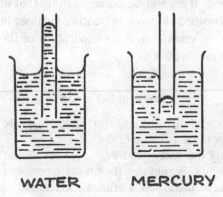

WATER   MERCURY

Fig. 20. Capillary Action

Capillary action has many applications. The subsurface water in fields is carried up to the roots of plants through tiny pores (capillary tubes) in the soil. A blotter works by capillary action. A towel consists of thousands of tiny capillary tubes which draw water from your skin to dry it. A sponge "drinks" water into its capillary tubes.

THE SOLID STATE

As you know, when liquids are sufficiently cooled, they will congeal to the solid state. The temperature at which a substance solidifies from the liquid to the solid state is known as the **freezing point** of the substance. As in the case of the boiling point, the atmospheric pressure affects the freezing point of a substance. Some solids like ice and bismuth expand on cooling. Most solids shrink on cooling. An increase in pressure will lower the freezing point of solids which expand on cooling. Ice melts under pressure. For solids which shrink on cooling, an increase in pressure will raise the freezing point. In general, the effect of pressure on freezing point is much less dramatic than its effect on the boiling point. Consequently, for most considerations we can ignore the pressure effect.

In the solid state, the particles of a substance, whether these particles are molecules or ions, are not completely rigid. They are free to vibrate within definite spatial limits and thus possess some kinetic energy. Theoretically, all particles possess some energy until the temperature is reduced to absolute zero.

VITREOUS SOLIDS

Structurally, there are two classes of solids. One type is known as the **vitreous,** or **amorphous,** type. In such solids there is a completely random arrangement of the particles. This lack of definite geometrical pattern results in specialized properties for this type. When broken with a hammer, these solids exhibit a curved fracture surface. When heated, they soften gradually and slowly transform to the liquid state with no clearly defined melting point. Glass is a typical example of this type of solid. There is some evidence to indicate that such solids are not true solids, but in reality are highly rigid and extremely viscous liquids.

CRYSTALLINE SOLIDS

The second type of solid is known as a **crystalline** solid. In this type the particles are arranged in definite geometric patterns. The result is that these solids exhibit definite physical properties in contrast with vitreous solids. When struck with a hammer, crystalline solids break along definite planes. When heated, they melt at a specific temperature. Figure 21 shows the six basic configurations of particles in crystals. These configurations are also known as **crystal lattices.** Most metals, solid hydrogen, and table salt all solidify in the cubic system. Sodium sulfide and indium metal crystallize in the tetragonal system. Sulfur exhibits orthorhombic crystals. Sugar forms monoclinic crystals. Boric acid is a member of the triclinic system. Magnesium, zinc, quartz, and solid oxygen all belong to the hexagonal system.

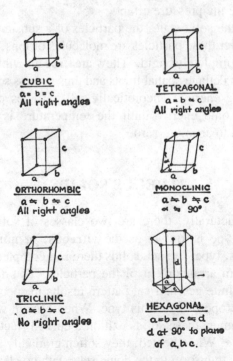

Fig. 21. The Basic Crystal Configurations

ENERGY AND CHANGE OF STATE

Relatively large amounts of energy are involved when a substance changes from one physical state to another. When a substance changes from gas to liquid to solid, energy is given up by the substance. A change from solid to liquid to gas involves the absorption of energy by the substance. The energy change at the gas-liquid transformation is known as **heat of vaporization.** The heat of vaporization of a substance is the number of calories required to change one gram of the liquid to a gas at the transformation temperature. The heat of vaporization of water at 100° C. is 540 cal./g. The energy change at the liquid-solid transformation is called the **heat of fusion.** The heat of fusion of a substance is the number of calories required to melt one gram of a solid to liquid at the transformation temperature. The heat of fusion of ice is 79.7 cal./g. at 0° C.

Thus, if we start with 1 gram of ice and intend to boil it, the energy requirements will be as follows:

To melt the ice:	79.7 cal.
To heat the water from 0° to 100°	100.0 cal.
To vaporize the water at 100°	540.0 cal.
Total	719.7 cal.

In the example just worked out, notice that 619.7 calories, or about 86% of the total energy requirements, were involved with changes of state. Only about 14% of the energy absorbed went into the raising of the temperature of the substance involved. It should be borne in mind that the energy absorbed or given off during changes in state causes no change in the temperature of the substance involved.

Problem Set No. 8

1. Which of the substances in Table XVIII are gases at room temperature (25° C.)?
2. Using the data found in Tables XVIII and XIX, compute the ratio of the absolute normal boiling point to the absolute critical temperature for each of the following:

(a) Carbon dioxide.
(b) Chlorine.
(c) Ethyl alcohol.
(d) Hydrogen.
(e) Oxygen.
(f) Water.

3. From your results in Problem 2, can you detect any relationship between these two properties of substances?

4. You are planning a trip by car to the top of Pike's Peak. Your car is equipped with a 180° thermostat (which prevents the circulation of water through your radiator below that temperature). Prove by calculations why it would be wise to remove the thermostat before making the ascent.

5. How many calories would be required to heat 25 grams of ice at 0° C. until it has completely boiled away?

SUMMARY

A gas can be condensed to a liquid by pressure only at temperatures below its critical temperature.

The **vapor pressure** of a liquid is the increase in pressure its vapor produces when it saturates the atmosphere above it in a closed container.

The **boiling point** of a liquid is the temperature at which its vapor pressure equals the atmospheric pressure about it.

Evaporation produces a cooling effect because only those molecules of highest energy escape from the liquid as vapor.

The **surface tension** in liquids is a result of attractive forces on surface molecules directed toward the interior of the liquid.

Cohesion is the attractive force between the molecules of a liquid. **Adhesion** is the attractive force between the molecules of a liquid and the molecules of its container.

Capillary action is the rise or depression of the surface of a liquid inside a small-diameter tube penetrating the surface.

The **freezing point** of a substance is the temperature at which it transforms from the liquid to the solid state.

Vitreous solids have random particle arrangement. **Crystalline solids** have a definite geometrical particle arrangement in one of six basic **crystal lattices.**

Heat of vaporization is the number of calories required to change one gram of a liquid to a vapor, at a given temperature.

Heat of fusion is the number of calories required to change one gram of a solid to a liquid at a given temperature.

CHAPTER 8

SOLUTIONS

A solution consists of two components, a **solvent** which is the dissolving medium, and a **solute** which is the substance dissolved. Solutions are mixtures because an infinite number of compositions involving a given solute and solvent are possible. In solutions the solute is dispersed into molecules or ions, and the distribution of the solute is perfectly homogeneous throughout the solution. A tremendous amount of chemistry takes place in solution, and so it is well for us to become thoroughly familiar with both the terminology and the properties of solutions.

METHODS OF EXPRESSING CONCENTRATION

A **concentrated solution** is one which contains a relatively large amount of solute per unit volume of solution. A **dilute solution** is one which contains a relatively small amount of solute per unit volume of solution. The words "strong" and "weak" should not be used when referring to the concentration of a solution, for these words have their own special meanings in chemistry and will be discussed later.

The principal methods of expressing the concentrations of solutions are:

1. **Molarity.** The molarity of a solution is the number of moles of solute per liter of solution. Molarity is abbreviated M. Its units are: No. of moles/liters. A solution which contains a half mole of solute per liter of solution is designated as 0.5 M. The following relationship is frequently useful in calculating the molarity of a solution.

$$\text{Molarity} = \frac{\text{No. of moles of solute}}{\text{Liters of solution}} \quad (1)$$

2. **Normality.** The normality of a solution is the number of equivalents of solute per liter of solution. Normality is abbreviated N. Its units are: No. of equivalents/liters. A solution containing 2.0 equivalents of solute per liter of solution is designated as 2.0 N. The following relationship is frequently useful in calculating the normality of a solution.

$$\text{Normality} = \frac{\text{No. of equivalents of solute}}{\text{Liters of solution}} \quad (2)$$

The number of equivalents of a substance is found from the relationship:

$$\text{No. of equivalents} = \frac{\text{Actual weight of subst.}}{\text{Equivalent weight}} \quad (3)$$

In Chapter 5 we saw that the **equivalent weight of an element** can be found from its atomic weight and its valence thus:

$$\text{Equivalent weight} = \frac{\text{Atomic weight}}{\text{Valence}} \quad (4)$$

Now we must learn to find the **equivalent weight of compounds.** To do this we begin with the formula of the compound, write it, and split the positive part of the formula from the negative as follows:

Na|Cl   K$_2$|CO$_3$ Ca|SO$_4$ Al|Cl$_3$ Fe$_2$|(SO$_4$)$_3$

We then consider the positive (left) part of the formula. The product of the valence of the positive part times its subscript gives us what is known as **the number of replaceable hydrogens** or **net positive valence** of the compound.

Net positive valence = (Valence of + element) × (Its subscript) (5)

The net positive valence of the compounds listed above would be:

NaCl: $1 \times 1 = 1$.   K$_2$CO$_3$: $1 \times 2 = 2$.
CaSO$_4$: $2 \times 1 = 2$.   AlCl$_3$: $3 \times 1 = 3$.
Fe$_2$(SO$_4$)$_3$: $3 \times 2 = 6$.

The equivalent weight of a compound is then found from the relationship:

$$\text{Equivalent weight} = \frac{\text{Molecular weight}}{\text{Net positive valence}} \quad (6)$$

The normality and molarity of a given solution are related as follows:

Normality = Molarity × Net positive valence (7)

Note that normality will be equal to or greater than the molarity. It is **never smaller** in magnitude.

3. **Molality.** The molality of a solution is the number of moles of solute per 1000 grams of solvent. Molality is abbreviated *m*. Its units are: No. of moles/1000 grams of solvent. Molarity and molality should not be confused. The former is defined in terms of the total volume of solution. The latter is defined in terms of a definite weight of solvent. A solution which contains 0.5 moles of solute in 250 grams of solvent would have a molality of:

$$0.5 \times \frac{1000}{250} = 2.0 \; m.$$

4. **Percentage Composition.** This may be either percentage by weight or percentage by volume.

$$\% \text{ by weight} = \frac{\text{Weight of solute}}{\text{Weight of solution}} \times 100. \quad (8)$$

$$\% \text{ by volume} = \frac{\text{Volume of solute}}{\text{Volume of solution}} \times 100. \quad (9)$$

Percentage by weight is usually used in referring to solids dissolved in liquids. Percentage by volume is normally used with reference to gases in gases or liquids in liquids.

STANDARD SOLUTIONS

A **standard solution** is any solution of accurately known concentration. Standard solutions may be made up in the following ways:

1. **Weight per Unit Volume.** In this method, a quantity of a pure chemical substance is accurately weighed and then dissolved in a quantity of the solvent. Then additional solvent is added until the total volume of solution is accurately known. From the weight of solute, the volume of solution, and either the molecular weight or the equivalent weight of the solute, the molarity

or the normality of the solution can readily be calculated. Let us look at an example.

EXAMPLE 1: 21.2 g. of Na_2CO_3 are dissolved in water and the solution is made up to 400 ml. Find the molarity and normality of the solution.

SOLUTION:

The molecular weight of Na_2CO_3 is 106.
Molarity is No. of moles/volume in liters.
No. of moles is: actual weight/molecular weight.
So:

$$\text{Molarity} = \frac{\text{actual weight/molecular weight}}{\text{volume in liters}}$$

$$= \frac{\text{actual weight}}{\text{molecular weight} \times \text{volume in liters}}$$

So, $\quad \text{Molarity} = \dfrac{21.2}{106 \times 0.400} = 0.5 \text{ M.}$

The net positive valence of Na_2CO_3 is: $1 \times 2 = 2$.

The equivalent weight of Na_2CO_3 is: $\dfrac{106}{2} = 53$.

The normality is: $0.5 \text{ M} \times 2 = 1.0 \text{ N.}$
Checking this result by working with the equivalent weight of Na_2CO_3, we have (by paralleling the working out of molarity)

$$\text{Normality} = \frac{\text{actual weight}}{\text{equivalent weight} \times \text{volume in liters}}$$

$$= \frac{21.2}{53 \times 0.400} = 1.0 \text{ N.}$$

2. **Dilution.** A definite volume of a more concentrated solution can be diluted with a definite amount of additional solvent to produce a more dilute solution of known concentration. The fact that **the number of moles or equivalents of solute does not change during dilution** enables us to calculate the new concentration. Note the following relationships.

Molarity × volume in liters = no. of moles of solute.
Molarity × volume in milliliters = no. of millimoles of solute.
Normality × vol. in liters = no. of equivalents of solute.
Normality × vol. in ml. = no. of milliequivalents of solute.

From these relationships we can see that when we are diluting solutions, the product of the concentration and volume of the initial solution must be equal to the product of the concentration and volume of the diluted solution when the same system of units is used in both solutions. Expressed as a relationship this would be:

$$C_i V_i = C_f V_f \qquad (10)$$

Where: C_i is concentration of initial solution.
V_i is volume of initial solution.
C_f is concentration of final solution.
V_f is volume of final solution.

Let us work out an example of this type of problem.

EXAMPLE 2: How much water must be added to 50 ml. of 1.2 M HCl solution in order to produce a 0.5 M HCl solution?

SOLUTION: Using the relationship:

$$C_i \times V_i = C_f \times V_f$$
$$1.2 \times 50 = 0.5 \times X$$
$$X = \frac{1.2 \times 50}{0.5} = 120 \text{ ml.}$$

120 ml. is the volume of the final solution. The amount of water to be added is then the difference between the volumes of the two solutions. So the volume of water to be added is:

$$120 - 50 = 70 \text{ ml. of water.}$$

3. **Reaction.** It is a fundamental law of chemistry that a given number of equivalents of one substance react with precisely the same number of equivalents of any other substance. Therefore if the concentration of a solution is unknown, its concentration can be found by measuring the volume of it which will react precisely with a definite weight of a pure solid substance, or with a definite volume of a standard solution. We have seen that the product of normality times volume in liters gives the number of equivalents of solute. Therefore the following relationship holds whenever solutions react:

$$N_1 \times V_1 = N_2 \times V_2 \qquad (11)$$

where: N_1 and V_1 are the normality and volume of one solution and N_2 and V_2 are the normality and volume of the other.

When a solution is reacting with a solid substance, then the following holds:

$$N_s \times V_s = \text{no. of equivalents of solid.} \qquad (12)$$

where: N_s and V_s are the normality and volume of the solution involved. **Here the volume must be in liters!**

Let us study some examples of these types.

EXAMPLE 3: 25.2 ml. of 0.1 N HCl solution are required to neutralize 20.0 ml. of an unknown base solution. Find the concentration of the base solution.

SOLUTION: Applying the relationship:

$$N_1 \times V_1 = N_2 \times V_2$$
$$0.1 \times 25.2 = X \times 20.0$$
$$X = \frac{0.1 \times 25.2}{20.0} = 0.126 \text{ N.}$$

EXAMPLE 4: 48.5 ml. of an unknown acid solution neutralize 11.1 g. of solid $Ca(OH)_2$. Compute the concentration of the acid solution.

SOLUTION: Applying the relationship:

$$N_s \times V_s \text{ (in liters)} = \text{no. of equivalents of solid}$$
$$X \times 0.0485 = \frac{11.1}{74/2}$$
$$X = \frac{2 \times 11.1}{74 \times 0.0485}$$
$$= 6.19 \text{ N.}$$

Notice that we initially had to determine the equivalent weight of $Ca(OH)_2$, which was its molecular weight, 74, divided by its net positive valence, 2.

Problem Set No. 9

1. Find the molarity of the following solutions.
    (a) 0.02 moles of NaOH in 80 ml. of solution.
    (b) 0.234 g. of NaCl in 50 ml. of solution.
    (c) 222 g. of $CaCl_2$ in 4 liters of solution.
2. Find the normality of the following solutions.
    (a) 0.2 moles of $CaCl_2$ in 200 ml. of solution.
    (b) 0.2 equivalents of $CaCl_2$ in 200 ml. of solution.
    (c) 2.76 g. of K_2CO_3 in 400 ml. of solution.
    (d) 6.84 g. of $Al_2(SO_4)_3$ in 250 ml. of solution.
3. Find the molality of the following solutions.
    (a) 4.0 g. of NaOH in 400 g. of water.
    (b) 333 g. of $CaCl_2$ in 6000 g. of water.
4. 6.0 g. of $MgSO_4$ are dissolved in 250 ml. of solution.
    (a) Find the percentage by weight of the solute (Specific gravity of water = 1).
    (b) Find the molarity of the solution.
    (c) Find the normality of the solution.

5. 10 ml. of ethyl alcohol are dissolved in 40 ml. of water. Find the percentage by volume of alcohol in the solution.

6. 34.0 g. of $AgNO_3$ are dissolved in 750 ml. of solution. Find the molarity of the solution.

7. 2.67 g. of $AlCl_3$ are dissolved in 400 ml. of solution. Find the normality of the solution.

8. How much water must be added to 35 ml. of 0.8 M $NaNO_3$ solution to make a 0.5 M solution?

9. 24.0 ml. of 0.1 N NaOH solution neutralize 25.0 ml. of an unknown acid solution. Compute the concentration of the acid solution.

10. 20.0 ml. of $AgNO_3$ solution precipitate 0.285 g. of AgCl from a salt water solution. Compute the normality of the $AgNO_3$ solution. (Remember, the number of equivalents of a substance formed is equal to the number of equivalents of each of the substances required to form it.)

SOLUBILITY

The solubility of a substance is the maximum amount of it that can be dissolved in a given amount of solvent at a specified temperature and pressure. The factors influencing solubility are:

1. Nature of solute and solvent
2. Temperature
3. Pressure

Let us consider each of these factors in terms of the common types of solutions. These types are: gas in a gas, gas in a liquid, liquid in a liquid, and solid in a liquid.

The first type, gas in a gas, has been discussed previously in relation to Dalton's Law of Partial Pressures and to gases in general. All gases are soluble in one another, and mix in all proportions. Molecules of one gas diffuse into the void between the molecules of the other and ultimately form a homogeneous system.

In the case of gas in a liquid, the temperature and pressure have important effects. When a gas is brought in contact with a liquid, some of the molecules of the gas will enter the liquid and dissolve in it. There appears to be no rule for predicting the solubility of a given gas in a given solvent. Gases like ammonia and hydrogen chloride are extremely soluble in water, whereas hy-

drogen and oxygen are only very slightly soluble in water. The effect of temperature in this case is quite definite. Just as an increase in temperature speeds up the rate of evaporation of a liquid, so an increase in temperature increases the rate at which dissolved gas molecules are expelled from the liquid. The rule is: **The solubility of any gas in a liquid solvent decreases as the temperature rises, and becomes zero at the boiling point of the solvent.**

EXPERIMENT 13: Select two bottles of some carbonated soda, one ice cold and one warm. Open the bottles, and with thumbs held over the tops, *gently* shake them. Which foams up more? Carbon dioxide gas has been dissolved under pressure in both. Its solubility is less in the warm sample, and therefore it escapes more easily from the warm soda. Hence the greater amount of foaming in the warm bottle.

The effect of pressure in this case is likewise quite definite. An increase in pressure on the gas will step up the rate at which gas molecules enter the liquid. The relationship between pressure and the solubility of a gas in a liquid is stated in **Henry's Law** thus: **For gases which do not react chemically with the solvent, the weight of gas which dissolves in a liquid at a given temperature is proportional to the partial pressure of the gas over the solution.**

In the case of liquids dissolving in liquids, there are three possibilities:

1. The two liquids are completely miscible (capable of being mixed).
2. The two liquids are partially miscible.
3. The two liquids are immiscible (not capable of being mixed).

In this case, the nature of the solute and the solvent has an important effect. The chemical structure of the substances involved determines to a great extent whether or not two liquids will dissolve in each other. Water and alcohol are soluble in each other in all proportions. Alcohol and carbon tetrachloride are soluble in each other in all proportions. But water and carbon tetrachloride are practically insoluble in each other. This strange behavior is better understood if we consider the structure of each of these substances.

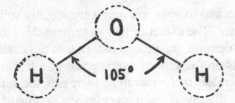

Fig. 22. The Water Molecule

The water molecule is a bent molecule as shown in Figure 22. The oxygen end of the molecule has a negative electrical character and the hydrogen end has a positive electrical nature. Thus, the **water molecule is highly polar** and possesses a high degree of electrical activity. On the other hand, carbon tetrachloride is made up of highly symmetrical molecules. In these molecules, the carbon atom is centrally located and the four chlorine atoms surround the carbon forming the apexes of a tetrahedron. This is shown in Figure 23. Such a molecule has very little polarity and electrically is virtually inert. The great difference in polarity between molecules of carbon tetrachloride and water accounts for the immiscibility of these two liquids.

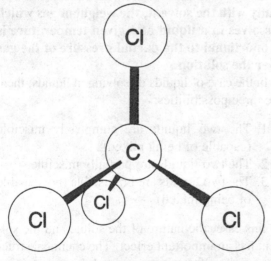

Fig. 23. Carbon Tetrachloride

The molecules of ethyl alcohol have a structure which produces a degree of polarity intermediate between water and carbon tetrachloride. This structure is illustrated in Figure 24. One end of the alcohol molecule has carbon atoms surrounded by hydrogen atoms, and is structurally

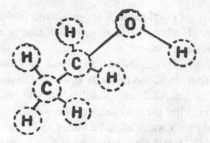

Fig. 24. Ethyl Alcohol

akin to the molecules of carbon tetrachloride. The other end of the molecule of alcohol is similar to the water molecule, and the lack of symmetry produces a moderate amount of polarity. The similarities in each case are sufficient to enable alcohol to dissolve in each of the other two liquids. The rule based upon this study is: **Like dissolves like.** "Like," of course, refers to the degree of polarity and the similarity of structural features of the molecules concerned.

A change in temperature changes the solubility of partially miscible liquids in one another, but the direction and magnitude of such change in solubility follows no general rule. Change in pressure has a negligible effect on the solubility of liquids in one another.

The most commonly encountered type of solution is, of course, the solid in a liquid. If a small amount of salt is added to a glass of water, we have a solution which is said to be **unsaturated** because more of the solute can be dissolved in the amount of solvent present. If we continue to add salt, we will ultimately reach the point where no further salt can be dissolved and the solution is said to be **saturated.** The concentration of a solute in a saturated solution is the solubility of the solute at the particular temperature. With some salts, like photographic hypo, sodium thiosulfate, $Na_2S_2O_3$, it is possible to form an unstable **supersaturated solution** which holds in solution more solute than can theoretically be held by the solvent at the particular temperature. Such solutions will revert back to saturated solutions by precipitating the excess solute as crystals if the supersaturated solution is shaken, or comes in contact with dust particles, or is "seeded" with a small crystal of the solute.

As a general rule, **the solubility of a solid in a liquid increases as the temperature rises.** There are, however, important exceptions to this general rule. Some salts exhibit a decrease in solubility with increased temperature. Such salts form scales in boilers and leave deposits in kettles and steam irons.

LAW OF PARTITION

If a solid is soluble in two liquid solvents which, in turn, are immiscible in each other, the solid will distribute itself between the solvents in quantities proportional to its solubility in each solvent. This is known as the **Law of Partition.** This phenomenon has application in the extraction of a solute from one solvent by another. For example, iodine is about 650 times more soluble in carbon tetrachloride than in water. Each time a sample of water containing dissolved iodine is shaken with carbon tetrachloride, the dissolved iodine will distribute itself between the solvents as follows: 650 parts by weight of iodine in carbon tetrachloride to 1 part by weight of iodine in water. Successive treatments of the water solution of iodine with fresh quantities of carbon tetrachloride will rapidly and effectively remove all appreciable traces of iodine from the water.

EFFECTS OF SOLUTES ON PROPERTIES OF SOLVENTS

Solutes may be divided into two classes:

1. Those which when dissolved in water produce a solution which conducts an electric current. Such solutes are called **electrolytes.** Acids, bases, and salts belong to this class. Solutions of electrolytes will be considered in the next chapter.

2. Those which when dissolved in water produce a solution not capable of conducting an electric current. Such solutes are called **nonelectrolytes.** Let us consider the effect of nonelectrolytes on the vapor pressure, freezing point, and boiling point of solvents.

LOWERING OF VAPOR PRESSURE

If a solute less volatile than the solvent is dissolved in the solvent, the vapor pressure of the solution will be lower than the vapor pressure of the solvent. You will recall that the pressure of the vapor of a liquid is the result of the escape of the vapor from the surface of the liquid into the air in contact with the liquid. You also know that a solute distributes itself uniformly throughout the solvent, and thus takes up a portion of the surface of the solution. This has the effect of decreasing the number of molecules of solvent in contact with the air above it, and consequently reduces the rate at which these molecules escape as vapor. This in turn lowers the vapor pressure.

The extent to which nonvolatile nonelectrolytes lower the vapor pressure of their solvents depends upon the concentration of the nonvolatile nonelectrolyte in the solution. **Raoult's Law** states the effect thus: **Equimolar quantities of different nonvolatile solutes, when added to equal weights of the same solvent, lower the vapor pressure the same amount, and the ratio of the amount of lowering to the vapor pressure of the pure solvent equals the ratio of the number of moles of solute to the number of moles of solution.** Raoult's Law may be put into mathematical form thus:

$$\frac{P - p}{P} = \frac{n}{N + n} \qquad (13)$$

where:

P = vapor pressure of solvent
p = vapor pressure of solution
N = number of moles of solvent
n = number of moles of solute.

In dilute solutions, where the number of moles of solvent is considerably larger than the number of moles of solute, the quantity (N + n) in Raoult's Law may be changed simply to (N) without causing too much error. This greatly simplifies calculations, and it modifies Raoult's Law to read:

$$\frac{P - p}{P} = \frac{n}{N} \qquad (14)$$

Of course, the number of moles of any substance

is its actual weight divided by its molecular weight. So, in the expression above,

$$n = \frac{w}{m} \quad \text{and} \quad N = \frac{W}{M}$$

where:

w and W are the actual weights of solute and solvent respectively, and
m and M are the molecular weights of solute and solvent respectively.

Substituting these into the expression above, and solving for m, the molecular weight of the solute, we obtain:

$$m = \frac{(w) \times (M) \times (P)}{(W) \times (P-p)} \qquad (15)$$

Thus, through the application of studies into the nature of vapor pressure, we arrive at a method for finding the molecular weight of solids of unknown composition. We simply find a suitable solvent for the solid, measure the actual weights of the two substances forming the solution, measure the vapor pressure of the solution, and then calculate the unknown molecular weight. This illustrates the familiar fact that investigation into one facet of the behavior of nature frequently leads to information revealing the secrets of nature in other areas.

ELEVATION OF BOILING POINT

The fact that a solution has a lower vapor pressure than the pure solvent neatly explains another phenomenon associated with solutions, namely that the boiling point of a solution is higher than the boiling point of the pure solvent. Examine Figure 25 carefully. Here the vapor pressure of a solvent has been plotted against temperature and is shown by the solid line. Such a curve would be obtained if we were to plot the data for water from Table XIV on page 43. The dotted curve shows how the vapor pressure curve would be depressed if a solute were added to the solvent. Now, a liquid boils when its vapor pressure equals the atmospheric pressure about it. The horizontal broken line in Figure 25 represents standard atmospheric pressure. Notice that the vapor pres-

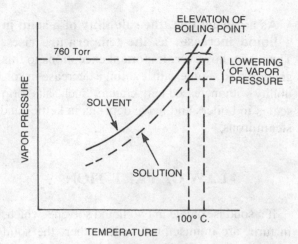

Fig. 25.
Effect of Solute
on Vapor Pressure Curve of a Liquid

sure curve of the solution reaches the atmospheric pressure line at a higher temperature than does the vapor pressure curve of the solvent. Thus the solution has a higher boiling point.

Investigations into this phenomenon have revealed that the amount of elevation of the boiling point for a given solution is directly proportional to the molality of the solution. Equimolal solutions of various solutes in the same solvent cause the same elevation in boiling point. For water, solutions of unit molality boil 0.52° C. higher than pure water at standard pressure. For water solutions, then, the following applies:

$$B - b = 0.52 \times m. \qquad (16)$$

where:

B is the boiling point of the solution
b is the boiling point of water
m is the molality of the solution.

It should be emphasized that the constant, 0.52, applies only to solutions in which water is the solvent. Each other solvent has its own constant which is, in each case, the number of degrees its boiling point is elevated when a nonvolatile solute is added to it to a concentration of 1 molal.

This relationship provides still another means of finding the molecular weight of a solute. Note the following expression:

Molality = no. of moles of solute/1000 g. of solvent.

Thus:

$$\text{Molality} = \frac{\text{Actual wt. of solute/1000 g. of solvent}}{\text{Molecular weight of solute}}$$

If we let:

w = actual weight of solute **per 1000 g. of solvent**
and m = molecular weight of solute,

and substitute this into Equation 16 above, we get:

$$B - b = \frac{0.52 \text{ w}}{m}. \qquad (17)$$

Solving this for m, the expression becomes:

$$m = \frac{0.52 \text{ w}}{B - b} \qquad (18)$$

Let us work through a problem of this type.

EXAMPLE 5: 42.75 g. of a substance are dissolved in 250 g. of water. The solution boils at 100.26° C. at standard pressure. Find the molecular weight of the solute.

SOLUTION:

$$w = 42.75 \times \frac{1000}{250} = 171 \text{ g.}$$

$$B - b = 100.26 - 100 = 0.26° \text{ C.}$$

Substituting into expression 18 we have:

$$m = \frac{0.52 \times 171}{0.26} = 342.$$

Thus the molecular weight of the substance is 342.

LOWERING OF FREEZING POINT

A solution freezes at a temperature below the freezing point of its solvent. This phenomenon is likewise related to vapor pressure. Examine Figure 26 carefully. You know that water boils at 100° and freezes at 0° at standard pressure. You are likewise aware that its boiling point decreases as the pressure drops and that its freezing point raises slowly as the pressure decreases. This means that as the pressure goes down, the range of temperature between the freezing and boiling points of water becomes smaller and smaller. Curve AB in Figure 26 shows how the boiling point changes as the pressure goes down, and curve AC shows the same for the freezing point.

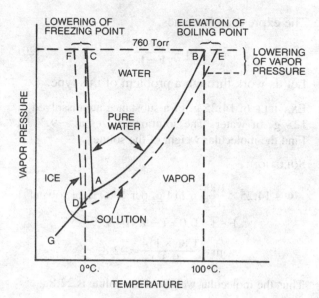

Fig. 26.
Temperature-Pressure Effects
on the Ice-Water-Vapor System
Caused by Solutes

The curves meet at point A, which is called the **triple point** because here ice, water, and vapor are all present at the same temperature and pressure. Curve AG shows the temperatures at which water vapor freezes directly to ice at pressures below the triple point.

As in the case of boiling point elevation, the amount of lowering of the freezing point is directly proportional to the molality of the solution. For water, solutions of nonelectrolytes of unit molality freeze at −1.86° C. For water solutions, then, the following applies:

$$F - f = 1.86 \times m. \qquad (19)$$

where:

F is the freezing point of water
f is the freezing point of the solution
m is the molality of the solution.

The constant 1.86 applies only to water. Every other solvent has its own particular constant which is the number of degrees a solution of unit molality freezes below the freezing point of the solvent.

Once again we have a method of finding the molecular weight of the solute. The derivation of an expression giving us the molecular weight of the solute from freezing point data proceeds exactly as in the case of boiling point elevation.

The expression becomes:

$$m = \frac{1.86\ w}{F - f}. \qquad (20)$$

Let us work through a problem of this type.

EXAMPLE 6: 14.25 g. of a substance are dissolved in 125 g. of water. The solution freezes at $-9.3°$ C. Find the molecular weight of the solute.

SOLUTION:

$$w = 14.25 \times \frac{1000}{125} = 114\ \text{g. per 1000 g. of solvent}$$

$$F - f = 0.0 - (-9.3) = 9.3°.$$

$$m = \frac{1.86 \times 114}{9.3} = 22.8.$$

Thus the molecular weight of the solute is 22.8.

APPLICATIONS OF EFFECTS OF SOLUTES

A variety of interesting and useful applications of the lowering of the vapor pressure of solvents by solutes may be pointed out at this time. For example, if a dish containing pure water and one containing a solution are placed side by side inside a tightly covered container, each liquid will begin to emit vapor into the air in the container. Since the vapor pressure of the solution is less than the vapor pressure of the water, the air in the container will become saturated with vapor relative to the solution first. Thus, any additional vapor emitted by the water will condense back to liquid **in the solution.** Consequently the air never gets a chance to become saturated with vapor relative to the water, with the result that all of the water eventually finds its way into the solution. This process is known as **isothermal distillation.**

EXPERIMENT 14: Select two 1-ounce glasses. Fill one half full of water. Place the same volume of saturated sugar water in the other. With a small piece of adhesive tape, mark the level of the liquid in each glass. Place the two glasses inside an airtight container, preferably of glass or clear plastic so you can observe any changes. Cover the container and set it aside in a location free from severe temperature changes. You will observe the sugar solution increase in volume at the expense of the water.

This phenomenon is the principle of operation of the various drying agents used in the cellar or in the closets of homes in moist climates. Solid chemicals, such as calcium chloride or potassium carbonate, which are very soluble in water, become moist in damp air. A saturated solution forms on their surface. The vapor pressure of this solution is less than the vapor pressure of the water in the moist air, and so more moisture is absorbed by the solution. Ultimately all of the solid dissolves in the moisture it absorbs, but even so, the solution continues to absorb moisture until the vapor pressure of the solution equals the pressure of the water in the atmosphere. This phenomenon is known as **deliquescence,** and deliquescent chemicals used as drying agents are called **desiccants.**

You are, of course, familiar with various applications of the lowering of freezing points. You have added alcohol or ethylene glycol (permanent antifreeze) to the radiator of your car to lower the freezing point of the water. We will learn in the next chapter why salt is more effective than nonelectrolytes such as sugar in causing the ice on your sidewalk to melt.

Problem Set No. 10

1. The vapor pressure of water at 20° is 17.5 torr and that of a solution of 23 g. of glycerin in 500 g. of water is 17.34 torr. Find the approximate molecular weight of the glycerin.
2. 12.4 g. of ethylene glycol, $C_2H_6O_2$, are dissolved in 200 g. of water. At what temperature will this solution boil at standard pressure?
3. 15.5 g. of a solute are dissolved in 100 g. of water. At standard pressure, the solution boils at 101.3° C. Find the molecular weight of the solute.
4. 11.5 g. of a solute are dissolved in 100 g. of water. The solution freezes at $-2.325°$ C. Find the molecular weight of the solute.
5. 6 quarts of ethylene glycol are mixed with 12 quarts of water in the radiator of a car. The specific gravity of the glycol is 1.26. Its molecular weight is 62. The specific gravity of water is 1. Find the Fahrenheit temperature at which this solution will freeze.

SUMMARY

A **solution** consists of a **solute** dissolved in a **solvent**. A **concentrated solution** contains more solute per unit volume than does a **dilute solution.**

A **standard solution** is one of accurately known concentration.

The **solubility** of a substance is the maximum amount of it that can be dissolved in a given amount of a solvent at a specified temperature and pressure.

For gases dissolved in liquids: as the temperature increases, the solubility decreases; as the pressure increases, the solubility increases.

For liquids dissolved in liquids: like dissolves like.

For solids dissolved in liquids: as the temperature increases, the solubility usually increases.

Nonvolatile **nonelectrolytes** alter the properties of their solvents as follows:

1. They lower the vapor pressure.
2. They elevate the boiling point.
3. They lower the freezing point.

Deliquescent solids dissolve in the moisture they absorb from the atmosphere.

CHAPTER 9

SOLUTIONS OF ELECTROLYTES

Electrolytes are solutes which, when dissolved in water, produce a solution which conducts an electric current. The behavior of solutions of electrolytes is remarkably different from the behavior of solutions of nonelectrolytes. Their chemistry is both interesting and important.

EXPERIMENT 15: Construct a conductivity apparatus as follows. Materials: 2 porcelain sockets, piece of wood approximately $6'' \times 4'' \times \frac{1}{2}''$, extension cord with male wall plug at one end, $5''$ piece of wire, 250 watt light bulb, 25 watt light bulb, clamp to hold the apparatus horizontal. Attach the two sockets side by side to the wood as shown in Figure 27. Wire the two sockets in series as shown in the same diagram. Remove the glass from the 250 watt bulb and also the thin filament between the filament support wires, and insert this into one of the sockets. Place the other bulb in the other socket. Clamp in place with the bulbs hanging down from the wood.

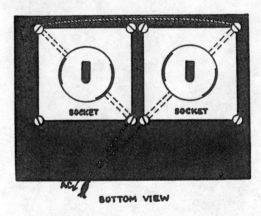

Fig. 27. Conductivity Apparatus

EXPERIMENT 16: Prepare water solutions of the following substances: sugar, alcohol, table salt, baking soda, washing soda, lye. Using the conductivity apparatus prepared in the previous experiment, check the conductivity of each of the solutions by lifting the container up to the wires in the opened light bulb. The 25 watt bulb will light if current passes between the immersed wires. **Use caution! Unplug the apparatus after each solution is tested,** and clean the wires with a cloth. Also check on the conductivity of pure water, pure alcohol, dry sugar, and dry table salt.

Which of the solutes are electrolytes? Must the solvent be present to have conductivity?

ABNORMAL BEHAVIOR OF ELECTROLYTES

Two major differences between the properties of electrolytes and nonelectrolytes are:

1. Solutions of electrolytes conduct electricity.
2. While electrolytes alter the properties of solvents in the same way as nonelectrolytes, they do so to a much greater degree.

Table XX presents data regarding the freezing points of solutions.

Table XX		
Freezing Points of Water Solutions		
Solute	*Molarity*	*Freezing Point in °C.*
Glycerin	0.1	−0.187
Ethyl Alcohol	0.1	−0.183
Sugar	0.1	−0.188
HCl	0.1	−0.352
KCl	0.1	−0.345
NaCl	0.1	−0.348
Na_2SO_4	0.1	−0.434
$CaCl_2$	0.1	−0.494
$NiCl_2$	0.1	−0.538

In Table XX, the first three substances are nonelectrolytes. The others are electrolytes. Note the difference in the lowering of the freezing point of water caused by the two types of solutes. The nonelectrolytes average 1.86° C. per mole of solute. HCl, KCl, and NaCl cause almost twice that amount of lowering, while Na_2SO_4, $CaCl_2$, and $NiCl_2$ produce more than twice and nearly three times as much lowering as nonelectrolytes. Boiling point elevation data and vapor pressure lowering data for these same solutes would show the same degree of dissimilarity. Such abnormal behavior requires explanation.

IONIZATION

The Theory of Ionization, first proposed by Arrhenius and then modified by subsequent investigation, adequately explains the behavior of dilute solutions of electrolytes. According to this theory, electrolytes dissociate into positively and negatively charged **ions** in solution. These charged ions are free to migrate through the solution, and thus are responsible for the conductivity of solutions of electrolytes.

The following equations show the dissociation of some electrolytes into their ions.

$$HCl = H^+ + Cl^-$$
$$NaCl = Na^+ + Cl^-$$
$$Na_2SO_4 = 2\ Na^+ + SO_4^{--}$$
$$CaCl_2 = Ca^{++} + 2\ Cl^-.$$

Note that **the charge on each ion is the same as the valence number of the atom or radical.**

Electrolytes may be divided into three types of substances:

1. **Acids.** Substances which ionize in solution to produce hydrogen ions (H^+).
2. **Bases.** Substances which ionize in solution to produce hydroxide ions (OH^-).
3. **Salts.** Substances which ionize in solution, but which produce neither hydrogen nor hydroxide ions.

The ionization equation of an electrolyte furnishes us with a strong clue concerning the explanation of the abnormal behavior of solutions of electrolytes. Notice that in the equation

$$KCl = K^+ + Cl^-$$

we see that for each mole of KCl that ionizes, one mole of potassium ions and one mole of chloride ions are formed. Thus we get two moles of particles in solution for each mole of solute dissolved, or twice Avogadro's number of particles in solution. A nonelectrolyte, which separates into **molecules** in solution, produces only one mole of particles in solution for each mole of solute dissolved. These facts lead directly to the idea that the alteration of properties of solvents by solutes depends not merely upon the concentration of the solute, but more precisely upon the total number of particles in solution. Since a salt like KCl produces twice as many particles in solution as substances like glycerin or sugar, we might expect that KCl would lower the freezing point of water twice as much as an equimolar solution of a nonelectrolyte. The data in Table XX are in general agreement with this idea. Actually, KCl lowers the freezing point of water about 1.85 times as much as the nonelectrolytes do. Similarly, $NiCl_2$, which ionizes

$$NiCl_2 = Ni^{++} + 2\ Cl^-$$

and which thus produces three moles of particles per mole of solute, lowers the freezing point of water almost three times as much as nonelectrolytes do. Actually $NiCl_2$ causes a lowering of about 2.9 times that of a nonelectrolyte.

The question might well be raised: Why doesn't KCl lower the freezing point of water precisely twice as much as a nonelectrolyte? The first answer given was: Electrolytes ionize slightly less than 100%. However, subsequent investigation caused this answer to be rejected. All of the electrolytes in Table XX ionize 100%. The answer lies, rather, in the fact that the ions formed possess electrical charges and are free to move about in the solution. As you know, opposite charges of electricity attract one another, so as the ions move about, they occasionally come close enough to an oppositely charged ion to be attracted to it and be held momentarily by it. This has the effect of producing a single particle from two particles. Now we know that water molecules are polar, and it is the polarity of water which causes ionic solutes to dissociate. The polar water molecules literally surround an ion, partially reduce the intensity of its attractive force, and float the ion off into the solution. When ions momentarily recombine, the water molecules quickly pull them apart. Nevertheless, at any one instant of time a percentage of the ions of an electrolyte will be momentarily held together. The number so held depends upon the concentration of the solution and upon the nature of the ions present. It will be larger in relatively more concentrated solutions, larger if the ions themselves are of relatively large size, and larger if the charge on the ions is more than one unit.

Let us look at the arithmetic involved as a result of these recombinations. Suppose that a solution initially contains 100 K^+ ions and 100 Cl^- ions. Let us suppose further that 10% of these ions are recombined at any instant. The total number of particles in solution will then be:

$$
\begin{array}{l}
90\ K^+\text{ions} \\
90\ Cl^-\text{ions} \\
\underline{10\ KCl\ \text{particles}} \\
190\ \text{total particles.}
\end{array}
$$

or,

Thus, the original 200 particles will be reduced in number by 5%. This explanation, based on the Theory of Ionization, explains the abnormal behavior of solutions of electrolytes.

CONCENTRATION OF IONS IN SOLUTION

When an electrolyte that ionizes 100% is dissolved in water, the concentration of the ions present depends upon the formula of the electrolyte dissolved. For example, if 0.1 mole of NaCl is dissolved in a liter of water, the concentration of each ion will be 0.1 M. A glance at the ionization equation of NaCl shows us why.

$$NaCl = Na^+ + Cl^-.$$

The equation tells us that for each mole of NaCl dissolved, one mole of each ion is formed. However, let us look at the ionization equation of salt like aluminum sulfate, $Al_2(SO_4)_3$.

$$Al_2(SO_4)_3 = 2\ Al^{+++} + 3\ SO_4^{--}.$$

Here, for each mole of salt dissolved, 2 moles of aluminum ion and 3 moles of sulfate ion are released into the solution. Thus, if 0.1 mole of aluminum sulfate is dissolved in a liter of water, the concentration of Al^{+++} will be 0.2 M, and the concentration of SO_4^{--} will be 0.3 M.

STRONG AND WEAK ELECTROLYTES

Ions possess electrical charges, and as a result, solutions containing ions are able to carry electrical currents. The conductivity of solutions of electrolytes has been thoroughly studied. Conductivity is, by definition, the reciprocal of electrical resistance. The units of conductivity are: **mho**, pronounced "reciprocal ohms"! The conductivities of many solutions have been accurately measured. We may assume that the total amount of electricity carried by a solution of an electrolyte is the sum of the electricity carried by each of the ions present. Table XXI gives the conductivity of 0.1 M solutions of the electrolytes indicated.

Table XXI
Conductivity of Solutions

Electrolyte	Molarity	Conductivity in mho
Acetic acid, $HC_2H_3O_2$	0.1	4.67
Hydrochloric acid, HCl	0.1	350.6
Nitric acid, HNO_3	0.1	346.4
Sodium acetate, $NaC_2H_3O_2$	0.1	61.9
Sodium chloride, NaCl	0.1	92.0
Sodium nitrate, $NaNO_3$	0.1	87.4

Let us now apply some arithmetical operations to these data to learn something about the nature of these electrolytes. We will use the letter C to indicate conductivity, and the symbol C_{HCl} will mean "the conductivity of HCl."

$$
\begin{array}{ll}
C_{HCl} = C_{H+} + C_{Cl-} & = 350.6\ \text{mho} \\
C_{NaNO_3} = C_{Na+} + C_{NO_3-} & = 87.4\ \text{mho}
\end{array}
$$

Adding these two equations we get:

$$C_{HCl} + C_{NaNO_3} = C_{H+} + C_{NO_3-} + C_{Na+} + C_{Cl-} = 438.0\ \text{mho}$$

Notice that the conductivity of each solution is fundamentally based upon the ability of ions to

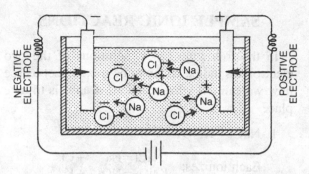

Fig. 28.
Electrical Conduction of Solutions of Electrolytes

carry electricity through the solution. Now suppose that we were to subtract from the above sum the conductivity of NaCl solution. This would have the effect of removing the influence of Na^+ and Cl^- **from both sides of the equation.** It would look like this:

$$C_{HCl} + C_{NaNO_3} = C_{H+} + C_{NO_3-} + C_{Na+} + C_{Cl-} = 438.0 \text{ mho}$$

$$C_{NaCl} = C_{Na+} + C_{Cl-} = 92.0 \text{ mho}$$

$$\overline{C_{HNO_3} = C_{H+} + C_{NO_3-} \qquad\qquad = 346.0 \text{ mho}}$$

This gives us a theoretical figure for the conductivity of HNO_3 solution.

The figure for nitric acid thus obtained is in excellent agreement with the figure in Table XXI. Let us now repeat this process to compute the conductivity of acetic acid, $HC_2H_3O_2$.

$$C_{HCl} = C_{H+} + C_{Cl-} \qquad = 350.6 \text{ mho}$$
$$C_{NaC_2H_3O_2} = C_{Na+} + C_{C_2H_3O_2-} \quad = 61.9 \text{ mho}$$

Adding, we get:

$$C_{HCl} + C_{NaC_2H_3O_2} = C_{H+} + C_{C_2H_3O_2-} + C_{Na+} + C_{Cl-} = 412.5 \text{ mho}$$
$$\text{But:} \quad C_{NaCl} = \qquad\qquad + C_{Na+} + C_{Cl-} = 92.0 \text{ mho}$$

Subtracting, we get:

$$C_{HC_2H_3O_2} = C_{H+} + C_{C_2H_3O_2-} = 320.5 \text{ mho}$$

This time our computed value for the conductivity of an electrolyte shows no agreement whatsoever with the measured value in Table XXI. The value for acetic acid in Table XXI is only about 1.4% of the computed value. Acetic acid and a few other electrolytes conduct a current of electricity only feebly. These facts lead to the idea that there are two different classes of electrolytes; one type known as strong electrolytes, and the other known as weak electrolytes.

A **strong electrolyte** is 100% ionized. A **weak electrolyte** is ionized only to a slight extent in solution. Always remember that the terms "strong" and "weak" refer to the degree of ionization of electrolytes, and never refer to their concentration.

EXPERIMENT 17: Compare the conductivity of table salt solution and a solution of baking soda with the conductivity of vinegar (acetic acid) and ammonia water (ammonium hydroxide). Which are strong and which are weak electrolytes?

The strong electrolytes are:

1. **The strong acids:** hydrochloric acid, HCl; sulfuric acid, H_2SO_4; and nitric acid, HNO_3. (A few less common acids such as HBr and HI are also strong.)
2. **The strong bases:** sodium hydroxide, NaOH; potassium hydroxide, KOH; magnesium hydroxide, $Mg(OH)_2$; and calcium hydroxide, $Ca(OH)_2$. (The latter two are only slightly soluble in water. In general, the hydroxides of the metals in Groups I and II of the periodic table are strong.)
3. **Practically all salts.**

REACTIONS OF ELECTROLYTES

The chemistry and properties of ions in solution are completely independent of the source of the ions. For example, Cu^{++} ions are blue in water solution whether they come from $CuCl_2$, $CuSO_4$, or $Cu(NO_3)_2$. Likewise, they will form a precipitate of $Cu(OH)_2$ in alkaline solutions regardless of their salt of origin. **Solutions of electrolytes are, in reality, solutions of the ions of the electrolyte.** Therefore the chemical reactions of electrolytes really are the reactions of the free ions in solution.

Ions in solution react with one another only under the following conditions:

1. If they can combine to form weak electrolytes.

2. If they can combine to form relatively insoluble substances.

3. If they oxidize or reduce one another or other molecules present in the solution.

The reactions of type 3 will be discussed in the next two chapters. We will concentrate at this time on the first two types of ionic reactions.

You now know all the weak electrolytes. Any electrolyte not included in the listing of strong electrolytes above may be considered as weak. Incidentally, **water is one of the most important weak electrolytes.** At room temperature it is ionized 0.00001%. As to the relatively insoluble substances, Table XXII will assist you in deciding whether or not a given electrolyte is soluble.

Table XXII
Solubility Chart

Substance	General Rules
Na^+	All sodium salts are **soluble.**
K^+	All potassium salts are **soluble.**
NH_4^+	All ammonium salts are **soluble.**
Ag^+	All silver salts, except $AgNO_3$, are **insoluble.**
$C_2H_3O_2^-$	All acetates, except silver, mercury, and bismuth acetate, are **soluble.**
NO_3^-	All nitrates are **soluble.**
Cl^-	All chlorides, except $AgCl$, $CuCl$, $PbCl_2$, and Hg_2Cl_2, are **soluble.**
SO_4^{--}	All sulfates except those of Ba, Pb, Ca, Hg, and Ag are **soluble.**
OH^-	All hydroxides except those of Na, K, NH_4, Ba, Ca, and Sr are **insoluble.**
CO_3^{--}	All carbonates except those of Na, K, and NH_4 are **insoluble.**
S^{--}	All sulfides except Na, K, NH_4, Mg, Ca, Ba, and Sr are **insoluble.**

SAMPLE IONIC REACTIONS

In the following cases, solutions of the two electrolytes indicated are to be mixed. Let us see how we can tell what reaction, if any, is to take place.

1. $NaCl + KNO_3$

Each ionizes:
$$NaCl = Na^+ + Cl^-$$
$$KNO_3 = NO_3^- + K^+$$

Examining the vertical arrangement of the ions present in the mixed solutions reveals that the potential products are both salts (strong electrolytes), and both are soluble. Therefore, **no reaction takes place in solution** in this case.

2. $NaCl + AgNO_3$

Each ionizes:
$$NaCl = Na^+ + Cl^-$$
$$AgNO_3 = NO_3^- + Ag^+$$

Examining the vertical arrangement of ions we see that the potential products are $NaNO_3$ and $AgCl$. Both are salts (strong electrolytes), but $AgCl$ is not soluble, so a reaction takes place in this case. The equation for the reaction may be written in the following forms:

(a) Formula form:
$$NaCl + AgNO_3 = \underline{AgCl} + NaNO_3.$$

The underline indicates that $AgCl$ precipitates out and that the remaining liquid is a solution of $NaNO_3$.

(b) Ionic form:
$$(Na^+,Cl^-) + (Ag^+,NO_3^-) = \underline{AgCl} + Na^+ + NO_3^-.$$

This indicates that a solid precipitates from two soluble electrolytes and shows the ions left in solution.

(c) Net ionic form:
$$Ag^+ + Cl^- = \underline{AgCl}.$$

This shows the heart of the reaction, indicating that the free ions **from any source** will undergo the reaction.

3. $Na_2CO_3 + HCl$

Each ionizes:

$$Na_2CO_3 = 2\,Na^+ + CO_3^{--}$$
$$2\,HCl = 2\,Cl^- + 2\,H^+$$

The vertical alignment indicates that one of the products is the weak acid, H_2CO_3, carbonic acid. Therefore, a reaction takes place. (Each of the substances in the second equation was given a coefficient of 2 in order to balance the charges in the two equations.) This reaction may then be written:

(a) Formula form:

$$Na_2CO_3 + 2\,HCl = H_2CO_3 + 2\,NaCl$$

There is no underline here because the H_2CO_3 remains in solution. (NOTE: Carbonic acid, H_2CO_3, is really a solution of the gas CO_2 in water. When CO_2 dissolves in water, it forms molecules of H_2CO_3 thus:

$$CO_2 + H_2O = H_2CO_3.$$

Now CO_2 is not highly soluble in water, so if the amount of H_2CO_3 formed in the reaction above is in excess of the amount that would be present in a saturated solution of CO_2, the H_2CO_3 will decompose and CO_2 will bubble out or effervesce from the solution, until all the excess CO_2 has been emitted. This same discussion applies to another weak acid, sulfurous acid, H_2SO_3, which is a solution of sulfur dioxide, SO_2, in water.)

(b) Ionic form:

$$(2\,Na^+, CO_3^{--}) + 2\,(H^+, Cl^-) = H_2CO_3 + 2\,Na^+ + 2\,Cl^-.$$

(c) Net ionic form:

$$2\,H^+ + CO_3^{--} = H_2CO_3.$$

4. $NaOH + HCl$

Each ionizes:
$$NaOH = Na^+ + OH^-$$
$$HCl = Cl^- + H^+$$

From the vertical alignment of ions, we see that one of the products is the weak electrolyte water, H_2O. So a reaction takes place.

(a) Formula form:

$$NaOH + HCl = H_2O + NaCl.$$

(b) Ionic form:

$$(Na^+, OH^-) + (H^+, Cl^-) = H_2O + Na^+ + Cl^-.$$

(c) Net ionic form:

$$H^+ + OH^- = H_2O.$$

This reaction between an acid and a base is known as **neutralization.** The net ionic form of the reaction tells us that **any acid, strong or weak, will neutralize any base, strong or weak!**

CHEMICAL EQUILIBRIUM

We have seen that weak electrolytes are only slightly, or partially, ionized. The conductivity data for acetic acid in Table XXI and the arithmetical treatment of it led us to that conclusion. But what accounts for the fact that every single 0.1 M solution of acetic acid at room temperature is ionized **to the same extent?** How do the molecules of acetic acid know when to stop ionizing? What keeps score for them on the extent of ionization?

Well, perhaps apologies are in order, for the last two questions are a bit unfair and misleading. Nature just doesn't behave that way. The explanation of the phenomenon of partial ionization can be found if we think about things we already know. When any electrolyte, strong or weak, is added to water, the polar water molecules begin to dissociate the electrolyte into its ions. Thus, a concentration of ions in solution begins to build up. The ions, of course, possess opposite electrical charges and thus can attract one another. In the case of strong electrolytes, water molecules are sufficiently polar to prevent any permanent recombinations of these ions. But in the case of weak electrolytes, water is less effective, and recombination of ions begins to take place as soon as any appreciable concentration of ions is present in the solution. So in the case of weak electrolytes we have **two processes taking place simultaneously in opposite directions,** dissociation and recombination. Initially, the dissociation takes place at a faster rate than the recombination, but eventually, as the concentration of ions builds up, the rate of recombination catches up to the rate of dissociation, after which, **both processes continue to proceed** at **the same rate.** The apparent effect of this is no change, for the two processes nullify each other. **When two opposing processes take place simultaneously at the same rate, a state of equilibrium exists.**

Actually we have seen several equilibrium situations in addition to partial ionization. When the air in contact with a liquid is saturated with the liquid's vapor, this means that the rate of evaporation is equal to the rate of condensation. When a solution is saturated with a solute, this means that the rate of dissolving is just equal to the rate of precipitation from solution. In each of these cases, both processes continue. The equality of rate creates the illusion of static conditions.

The **position of equilibrium** is the extent to which one of the processes progresses before the opposite process catches up with it. For example, a 0.01 M solution of acetic acid at room temperature is ionized about 4%. This means that if 0.01 mole of pure acetic acid is dissolved in a liter of water, about 4% of the solute will have ionized by the time equilibrium is reached. It also means that if 0.01 mole of hydrogen ion and 0.01 mole of acetate ion are added to a liter of water from **different** sources, the two ions will have combined to an extent of about 96% by the time equilibrium is reached.

The position of equilibrium is not fixed. It depends upon:

1. The nature of the substances involved.
2. The temperature.
3. The pressure (when gases are involved).
4. The concentration.

For example, each different weak electrolyte has its own degree of ionization at a given temperature, pressure, and concentration. An increase in temperature speeds up a process which absorbs energy and slows down one that gives off energy. Thus an increase in temperature will shift the position of equilibrium in the direction of the process absorbing energy. In many reactions involving gases, the number of moles of gas on each side of the equation is not necessarily the same. A difference in number of moles of gas is, of course, a difference in volume. An increase in pressure always shifts the position of equilibrium in the direction of the reaction producing the smaller volume. The **Principle of Le Chatelier** sums up the effects of changes in any of the factors influencing the position of equilibrium. It states: **A system in equilibrium, when subjected to a stress resulting from a change in temperature, pressure, or concentration, and causing the equilibrium to be upset, will adjust its position of equilibrium to relieve the stress, and reestablish equilibrium.**

LAW OF MASS ACTION

The effect of concentration on equilibrium is contained in the **Law of Mass Action.** This law states: **The velocity of a reaction is proportional to the product of the molar concentrations of the reacting substances, taken to proper powers, the powers being the coefficients of the reactants in the balanced equation for the reaction.** Consider the following hypothetical reaction:

$$mA + nB = pC + qD.$$

Let us assume that an equilibrium will be established between the substances on the left and those on the right. This means that we are thus dealing with two reactions, one proceeding to the right and the other proceeding to the left. According to the Law of Mass Action, the rate of the reaction to the right is:

$$r_1 = k_1 \times (A)^m \times (B)^n$$

where:

   r_1 is the reaction velocity or rate,
  (A) is the molar concentration of substance A,
  (B) is the molar concentration of substance B,
   k_1 is the constant of proportionality.

Note with care the use of parentheses to indicate the **molarity.**

Similarly, by the Law of Mass Action, the velocity or rate of the reaction to the left will be:

$$r_2 = k_2 \times (C)^p \times (D)^q.$$

Now, at equilibrium, the two rates will be equal. So, if

$$r_2 = r_1 \qquad \text{then:}$$
$$k_2 \times (C)^p \times (D)^q = k_1 \times (A)^m \times (B)^n$$
$$\frac{(C)^p \times (D)^q}{(A)^m \times (B)^n} = \frac{k_1}{k_2} = K. \qquad (1)$$

Equation 1 is known as the **generalized Law of Mass Action expression.** It applies to all equilibrium situations. Note that the numerator contains the molar concentrations of the substances, to proper powers, which usually appear on the right side in a balanced chemical equation. This is an adopted convention. The constant, K, is a constant for a given temperature. Let us now examine some of the applications of the Law of Mass Action expression to weak electrolyte systems.

IONIZATION CONSTANT

If we know the percentage ionization of a weak electrolyte and its concentration, we can compute K in the Law of Mass Action expression for the electrolyte. In this case, K is known as the **ionization constant.**

EXAMPLE 1: A 0.1 M solution of acetic acid, $HC_2H_3O_2$, is 1.32% ionized at equilibrium. Find its ionization constant.

SOLUTION: Acetic acid ionizes:

$$HC_2H_3O_2 = H^+ + C_2H_3O_2^-$$

The Law of Mass Action expression for this reaction is:

$$\frac{(H^+) \times (C_2H_3O_2^-)}{(HC_2H_3O_2)} = K.$$

We are given that 1.32% of the acid is in ionic form.

Therefore: $(H^+) = 0.1 \times 0.0132 = 0.00132$ M

$(C_2H_3O_2^-) =$ the same, for the ions are formed in equal amount.
$(HC_2H_3O_2) = 0.1 - 0.00132 = 0.09868$ M

Substituting into the expression, we obtain:

$$\frac{(0.00132)^2}{0.09868} = 1.77 \times 10^{-5}.$$

This constant applies to all equilibrium solutions of acetic acid at room temperature.

NOTE: In finding $(HC_2H_3O_2)$, we subtracted a very small number from a relatively larger one. If we had ignored this subtraction, our answer would not have been materially affected. Therefore we will ignore the effect of that part of the weak electrolyte which has ionized in future applications. We will assume, instead, that the concentration of electrolyte remaining un-ionized is the same as the original concentration of the electrolyte.

Once we know the ionization constant for a weak electrolyte, we are then able to calculate additional properties of other solutions of this electrolyte.

EXAMPLE 2: The ionization constant of acetic acid is 1.8×10^{-5}. Find the (H^+) in a 0.01 M solution of this acid.

SOLUTION: First we write the ionization equation:

$$HC_2H_3O_2 = H^+ + C_2H_3O_2^-.$$

From this, we write the Law of Mass Action expression:

$$\frac{(H^+) \times (C_2H_3O_2^-)}{(HC_2H_3O_2)} = 1.8 \times 10^{-5}.$$

Now, let $(H^+) = X$.
Then $(C_2H_3O_2^-) = X$, because both are formed in the same amount.

$$(HC_2H_3O_2) = 0.01.$$

Substituting, we get,

$$\frac{X^2}{0.01} = 1.8 \times 10^{-5}$$
$$X^2 - 1.8 \times 10^{-7} = 18 \times 10^{-8}$$
$$X = 4.2 \times 10^{-4} \text{ M} = 0.00042 \text{ M}.$$

EXAMPLE 3: What is the percentage of ionization of the acid in Example 2?

SOLUTION: If the 0.01 M acetic acid were 100% ionized, the concentration of H^+ would be 0.01 M. So to find the percentage ionization of the acid, we divide the actual (H^+) by 0.01 and multiply the result by 100, thus:

$$\frac{4.2 \times 10^{-4} \times 10^2}{1 \times 10^{-2}} = 4.2\%.$$

EXPERIMENT 18: Compare the percentage ionization of 0.1 M and 0.01 M acetic acid in Examples 1 and 3 above. Check the conductivity of pure vinegar in the conductivity apparatus made in Experiment 15. Now add small quantities of water (is water a conductor?) to the vinegar, checking the conductivity after each addition. You will observe the gradual brightening of the lamp because the percentage ionization of a weak electrolyte increases with dilution. Repeat the procedure with ammonia water.

pH

The concentration of hydrogen ions in any water solution is normally a matter of importance, for the hydrogen ion is responsible for all acid properties. Water is a very weak electrolyte, which means that it ionizes to a slight extent as follows:

$$H_2O = H^+ + OH^-.$$

Actually, H^+ ion is hydrated. This fact is indicated by writing $H^+(H_2O)$ or, more usually, H_3O^+. H_3O^+ is called the **hydronium ion.** The acidic properties of water are ascribed to H^+ or H_3O^+. In this text we will write H^+, keeping in mind that it represents H_3O^+. Note also that the dissociation of water can be written:

$$2 H_2O = H_3O^+ + OH^-.$$

Thus any time water is present, some of its ions are likewise present. It is known that water is 0.00001% ionized at room temperature. Therefore the concentration of each of the ions in pure water at room temperature is 1×10^{-7} M.

As you may have observed, the concentration of hydrogen ion in pure water and in dilute solutions of weak acids is usually a very small number. To avoid the use of such numbers, scientists have devised a scale for indicating the concentration of hydrogen ion which is known as the pH scale. It is defined as follows: **the pH is the logarithm of the reciprocal of the molar concentration of the hydrogen ion.** In mathematical form, it is:

$$pH = \log \frac{1}{(H^+)}. \qquad (2)$$

The pOH is defined in similar terms. Mathematically, it is:

$$pOH = \log \frac{1}{(OH^-)}. \qquad (3)$$

The pH of pure water may now be computed.

$$pH = \log \frac{1}{1 \times 10^{-7}} = \log 10^7 = 7.$$

Since, in pure water, the $(OH^-) = (H^+)$, the pOH will also be 7. From the Law of Mass Action, it can be shown that the following important relationship holds whenever water is present:

$$(H^+) \times (OH^-) = K_w = 1 \times 10^{-14} \qquad (4)$$

Therefore, for any system containing water:

$$pH + pOH = 14. \qquad (5)$$

We have seen that the pH of pure water is 7. If a solution has a **pH of less than 7,** it means that the concentration of hydrogen ion is **greater** than it is in pure water. Therefore, such solutions exhibit **acidic** properties. Similarly, if the pH is **greater than 7,** the solution will be **basic.**

Table XXIII gives the logarithms of numbers from 1.0 to 9.9. It will be of value in finding the pH of solutions in problems.

Let us now study a few examples involving the calculation of pH.

EXAMPLE 4: A 0.1 M solution of acetic acid has a hydrogen ion concentration of 1.3×10^{-3} M. Find the pH of the solution.

SOLUTION:

$$pH = \log \frac{1}{(H^+)}$$

$$= \log \frac{1}{1.3 \times 10^{-3}}$$

$$= \log \frac{10^3}{1.3} = \log 10^3 - \log 1.3$$

$$= 3 - \log 1.3.$$

$$pH = 3 - 0.11 = 2.89.$$

Note the make-up of the second to the last line in the solution. It is the exponent of 10 (with sign changed) minus the log of the coefficient of 10. This can be taken as a rule for writing pH quickly. Thus if a $(H^+) = 2.8 \times 10^{-4}$, the pH would be $4 - \log 2.8$. The

Table XXIII

Logarithms of Numbers

N	log	N	log	N	log	N	log	N	log
1.0	0.000	3.0	0.477	5.0	0.699	7.0	0.845	9.0	0.954
1.1	0.041	3.1	0.491	5.1	0.708	7.1	0.851	9.1	0.959
1.2	0.079	3.2	0.505	5.2	0.716	7.2	0.857	9.2	0.964
1.3	0.114	3.3	0.519	5.3	0.724	7.3	0.863	9.3	0.968
1.4	0.146	3.4	0.531	5.4	0.732	7.4	0.869	9.4	0.973
1.5	0.176	3.5	0.544	5.5	0.740	7.5	0.875	9.5	0.978
1.6	0.204	3.6	0.556	5.6	0.748	7.6	0.881	9.6	0.982
1.7	0.230	3.7	0.568	5.7	0.756	7.7	0.886	9.7	0.987
1.8	0.255	3.8	0.580	5.8	0.763	7.8	0.892	9.8	0.991
1.9	0.279	3.9	0.591	5.9	0.771	7.9	0.898	9.9	0.996
2.0	0.301	4.0	0.602	6.0	0.778	8.0	0.903		
2.1	0.322	4.1	0.613	6.1	0.785	8.1	0.908		
2.2	0.342	4.2	0.623	6.2	0.792	8.2	0.914		
2.3	0.362	4.3	0.633	6.3	0.799	8.3	0.919		
2.4	0.380	4.4	0.643	6.4	0.806	8.4	0.924		
2.5	0.398	4.5	0.653	6.5	0.813	8.5	0.929		
2.6	0.415	4.6	0.663	6.6	0.820	8.6	0.934		
2.7	0.431	4.7	0.672	6.7	0.826	8.7	0.940		
2.8	0.447	4.8	0.681	6.8	0.833	8.8	0.944		
2.9	0.462	4.9	0.690	6.9	0.839	8.9	0.949		

pH of 2.89 in this example is less than 7. This indicates that the solution is definitely acidic.

EXAMPLE 5: The ionization constant of ammonium hydroxide, NH_4OH, is 1.8×10^{-5}. For a 0.01 M solution of this substance, find:

(a) The concentration of OH^-.
(b) The pH.
(c) The percentage ionization.

SOLUTION:
(a) The ionization equation is:

$$NH_4OH = NH_4^+ + OH^-.$$

The Law of Mass Action expression, then, is:

$$\frac{(NH_4^+) \times (OH^-)}{(NH_4OH)} = 1.8 \times 10^{-5}$$

Let:

$$(OH^-) = X.$$

Then:

$$(NH_4^+) = X, \text{ (both formed in same amount)}$$
$$(NH_4OH) = 0.01.$$

Substituting into the Law of Mass Action expression:

$$\frac{X^2}{0.01} = 1.8 \times 10^{-5}$$

$$X^2 = 1.8 \times 10^{-7} = 18 \times 10^{-8}.$$

So,

$$(OH^-) = X = 4.2 \times 10^{-4} \text{ M}.$$

(b)

$$pOH = \log \frac{1}{(OH^-)}$$

$$= 4 - \log 4.2 = 4 - 0.62 = 3.38.$$

$$pH = 14 - pOH = 14 - 3.38 = 10.62.$$

Since the pH is greater than 7, the solution is basic.

(c)

$$\text{Percentage ionization} = \frac{\text{conc. of ions}}{\text{conc. of electrolyte}} \times 100.$$

$$= \frac{4.2 \times 10^{-4} \times 10^2}{1 \times 10^{-2}} = 4.2\%.$$

COMMON ION EFFECT

An interesting phenomenon occurs when a common ion is added to a solution of a weak electrolyte. Consider the ionization of ammonium hydroxide: $NH_4OH = NH_4^+ + OH^-$. Suppose that more ammonium ion were to be added to a solution of NH_4OH by dissolving some NH_4Cl salt in the solution. According to the rule of Le Chatelier, the increase in concentration of ammonium ion would upset the equilibrium in the equation above, and the position of equilibrium would shift toward the left to relieve the stress. This, then, should have the effect of decreasing the concentration of hydroxide ions in solution. In other words, the addition of the salt would have the effect of partially neutralizing the base! Let us check this with a numerical example.

EXAMPLE 6: To a 0.01 M solution of NH_4OH, sufficient NH_4Cl is added to increase the concentration of NH_4^+ to 0.03 M. Find:

(a) The concentration of OH^- in the solution.
(b) The pH of the solution.

SOLUTION: The ionization equation is:

$$NH_4OH = NH_4^+ + OH^-$$

The Law of Mass Action expression is:

$$\frac{(NH_4^+) \times (OH^-)}{(NH_4OH)} = 1.8 \times 10^{-5}.$$

(a) Let: $(OH^-) = X$.
$(NH_4^+) = 0.03$ M
$(NH_4OH) = 0.01$ M

Substituting into the Law of Mass Action expression:

$$\frac{0.03 \times X}{0.01} = 1.8 \times 10^{-5}.$$

$(OH^-) = X = 0.6 \times 10^{-5} = 6.0 \times 10^{-6}$ M.

In Example 5, we saw that the (OH^-) in a 0.01 M solution of NH_4OH was 4.2×10^{-4}. So the effect of adding the common ion was indeed to lower the concentration of hydroxide ion.

(b) $pOH = 6 - \log 6.0 = 6 - 0.78 = 5.22$.
$pH = 14 - pOH = 14 - 5.22 = 8.78$.

The addition of the ammonium ion changed the pH from 10.62 to 8.78, a value much closer to 7. So we see that the salt actually partially neutralizes the base.

The common ion effect may be summarized as follows: **The addition of an ion in common with an ion of a solute represses the dissociation of the solute.**

HYDROLYSIS

When an acid and a base neutralize each other, the products formed are a salt and water. On this basis, four different types of salts are possible:

1. A salt of a strong acid and a strong base, e.g., NaCl from NaOH and HCl.
2. A salt of a strong acid and a weak base, e.g., NH_4Cl from NH_4OH and HCl.
3. A salt of a weak acid and a strong base, e.g., Na_2CO_3 from NaOH and H_2CO_3.
4. A salt of a weak acid and a weak base, e.g., NH_4CN from NH_4OH and HCN.

Salts may therefore possess ions of weak electrolytes which are capable of reacting with the ions of the water in which the salt is dissolved to form molecules of weak electrolytes in solution. Such a reaction is known as **hydrolysis.** Let us investigate each type of salt to see what the possibilities for hydrolysis might be.

CASE 1: Salt of a strong base and a strong acid.

NaCl ionizes: $NaCl = Na^+ + Cl^-$

Its solvent, H_2O, ionizes: $H_2O = OH^- + H^+$

Examination of the vertical arrangement of the ions formed in solution shows that the potential products in each case, NaOH and HCl, are both strong electrolytes, both are 100% ionized, and thus neither will form molecules in solution. Therefore this class of salt does not hydrolyze in solution.

CASE 2: Salt of a weak base and a strong acid.

NH$_4$Cl ionizes: $\qquad$ NH$_4$Cl $=$ NH$_4$$^+$ $+$ Cl$^-$

Its solvent, H$_2$O, ionizes: $\qquad$ H$_2$O $=$ OH$^-$ $+$ H$^+$

Examination of the vertical alignment of ions present quickly shows that the weak electrolyte, NH$_4$OH, will form in this solution. Note that this will sweep hydroxide ions out of solution and bind them up in the NH$_4$OH molecules. Hydrogen ions will thus be left in excess in the solution, and the solution will become acidic. Therefore this class of salt hydrolyzes to produce acidic solutions.

CASE 3: Salt of a strong base and a weak acid.

Na$_2$CO$_3$ ionizes: $\qquad$ Na$_2$CO$_3$ $=$ 2 Na$^+$ $+$ CO$_3$$^{--}$

Its solvent, H$_2$O, ionizes: $\qquad$ 2 H$_2$O $=$ 2 OH$^-$ $+$ 2 H$^+$

Examination of the vertical alignment of the ions present quickly shows that the weak electrolyte, H$_2$CO$_3$, carbonic acid, will form in this solution. Note that this will sweep hydrogen ions out of solution and bind them up in molecules of carbonic acid. Hydroxide ions will thus be left in excess in solution, and the solution will become basic. Therefore this class of salt hydrolyzes to produce basic solutions.

CASE 4: Salt of a weak base and a weak acid.

NH$_4$CN, ammonium cyanide, ionizes:
$$NH_4CN = NH_4^+ + CN^-$$

Its solvent, H$_2$O, ionizes: $\qquad$ H$_2$O $=$ OH$^-$ $+$ H$^+$

Examination of the vertical alignment of ions present shows that two weak electrolytes, ammonium hydroxide and hydrocyanic acid, will form in this solution. Most of the ions will be swept out of this solution and it will thus be a very poor conductor of electricity. The acidity or alkalinity of this solution depends upon which of the two weak electrolytes formed is weaker. In this case, the ionization constant of NH$_4$OH is 1.8×10^{-5}, while the ionization constant of HCN is 4×10^{-10}. This shows that HCN is a much weaker electrolyte than NH$_4$OH, and that HCN is ionized to a much less extent. Therefore more hydrogen ions will be removed from this solution than hydroxide ions, and the solution will become very slightly basic. Salts of this class hydrolyze extensively producing solutions which are

either slightly acidic or slightly basic, depending upon the particular weak electrolytes formed in the solution.

EXPERIMENT 19: Prepare water solutions of table salt, NaCl, baking soda, NaHCO$_3$, and washing soda, Na$_2$CO$_3$. To each of these salt solutions add a few drops of phenolphthalein solution prepared in Experiment 5. In which 2 solutions is hydrolysis taking place? Compare them for degree of hydrolysis. You will observe that the washing soda is considerably more alkaline.

There are two other conditions under which the hydrolysis of a salt will take place:

1. If an insoluble hydroxide forms.
2. If a volatile gas forms.

Let us briefly look at these possibilities.

Magnesium chloride, MgCl$_2$, ionizes:
$$MgCl_2 = Mg^{++} + 2 Cl^-$$

Its solvent, H$_2$O, ionizes: 2 H$_2$O $=$ 2 OH$^-$ $+$ 2 H$^+$

Examination of the vertical alignment of ions present shows that Mg(OH)$_2$, a strong base but an insoluble one, will form in this solution. This will remove hydroxide ions from solution, leave hydrogen ions behind in excess, and thus produce an acid solution.

We have seen previously that NaCl does not hydrolyze in solution. However, if a solution of NaCl is vigorously boiled, HCl, a volatile gas, will be expelled from this solution. Thus hydrogen ions will be removed from the solution, and the solution will become basic as a result of the build-up of hydroxide ions in the solution.

Hydrolysis is an extremely important phenomenon and has many applications in industry. For example, the water in the boilers of factories or ships is made alkaline by adding sodium carbonate to it rather than sodium hydroxide in most cases. If acid is spilled on the skin, a solution of sodium carbonate or sodium bicarbonate effectively neutralizes the acid.

SOLUBILITY PRODUCT

We have a tendency to estimate the solubility of a substance by eye. For example, if we add some solute to water, stir the solution, and the solute disappears, we say that the solute was soluble. If, on the other hand, the solute does not disappear, we say that it is insoluble. Actually, when we use the word "insoluble" in chemistry, we mean "slightly soluble," for all electrolytes are soluble to some extent in water. However, many of them are so slightly soluble that we cannot detect by eye any dissolving taking place. These slightly soluble substances play an important part in analytical chemistry, a field in which we measure the composition of a substance or a mixture.

A saturated solution of an electrolyte has some special properties. As you know, in a saturated solution a state of equilibrium exists between the ions in solution and the ions in the crystals of the solute. All the rules of equilibrium thus apply to saturated solutions.

EXPERIMENT 20: Taste both table salt and baking soda (NaCl and NaHCO₃). Prepare a saturated solution of table salt. Add a quarter teaspoon of baking soda. Allow the precipitate to settle. After carefully pouring off the excess liquid, filter the solution through a handkerchief to collect the solid precipitate. Permit the solid in the handkerchief to dry. Taste it. What is it? You will observe the effect of a common ion, the sodium ion, on a saturated solution.

Furthermore, a law similar to the Law of Mass Action applies to saturated solutions of slightly soluble electrolytes. This law is the **Solubility Product Principle,** and it states: **At saturation, the product of the molar concentrations of the ions of the electrolyte, raised to proper powers, is a constant.** The powers are the coefficients in the ionization equation of the electrolyte. The constant is constant at a particular temperature. Thus, for the hypothetical electrolyte A_mB_n, its ionization equation would be: $A_mB_n = mA + nB$. Its solubility product expression would then be:

$$(A)^m \times (B)^n = K_{sp}. \qquad (6)$$

where the parentheses again indicate **the molar concentration of the substances included between them.** We can find the solubility product constant, K_{sp}, of a given electrolyte if we know its solubility in water.

EXAMPLE 7: The solubility of calcium fluoride, CaF_2, is 0.0016 g. per 100 ml. of water. Find its solubility product constant.

SOLUTION: The molecular weight of CaF_2 is 78. Its solubility in grams per liter is 0.016. Its solubility in moles per liter is $0.016/78 = 2.05 \times 10^{-4}$ M. CaF_2 ionizes: $\qquad CaF_2 = Ca^{++} + 2 F^-$

Its solubility product expression, therefore, is:

$$(Ca^{++}) \times (F^-)^2 = K_{sp}.$$

To find the constant we proceed in two steps.

1. Find the molar concentration of each of the ions.
2. Substitute these concentrations into the expression **and do with them what the expression tells us to do.**

The ionization equation tells us that for each mole of CaF_2 that dissolves, 1 mole of calcium ion and 2 moles of fluoride ion are released into the solution. So, (Ca^{++}) in solution is 2.05×10^{-4} M. (F^-) in solution is $2 \times 2.05 \times 10^{-4} = 4.1 \times 10^{-4}$ M. Substituting into the expression, we obtain:

$$(2.05 \times 10^{-4}) \times (4.1 \times 10^{-4})^2 = K_{sp}.$$
$$(2.05 \times 10^{-4}) \times (16.8 \times 10^{-8}) = K_{sp}.$$
$$K_{sp} = 3.4 \times 10^{-11}$$

The solubility product constant for slightly soluble substances, like the ionization constant of weak electrolytes, is a property of the substance. It applies to all saturated solutions of the substance. If we know the solubility product constant of a substance, we can find the concentration of the ions in a saturated solution through the use of the solubility product expression.

EXAMPLE 8: The solubility product constant of silver chloride is 1.6×10^{-10}. Find the concentration of the ions in a saturated solution of this salt.

SOLUTION: $\qquad$ AgCl ionizes: $\qquad AgCl = Ag^+ + Cl^-$

Its solubility product expression is:

$$(Ag^+) \times (Cl^-) = 1.6 \times 10^{-10}.$$

Let $X = (Ag^+)$. Then $X = (Cl^-)$ also, for the two ions are formed in equal amounts according to the ionization equation.

Substituting into the solubility product expression:

$$X^2 = 1.6 \times 10^{-10}$$
$$(Ag^+) = (Cl^-) = X = 1.3 \times 10^{-5} \text{ M}.$$

The following example can be very revealing as

to the nature of the solubility product principle and the behavior of solutions of ions of slightly soluble substances. Study it carefully.

EXAMPLE 9: A solution contains both chloride ion, Cl^-, and chromate ion, CrO_4^{--}. The (Cl^-) is 0.001 M, and the (CrO_4^{--}) is 0.005 M. Silver ion, in the form of a solution of silver nitrate, is to be added slowly to this solution. The K_{sp} of AgCl is 1.6×10^{-10}. The K_{sp} of Ag_2CrO_4 is 9.0×10^{-12}.

(a) At what (Ag^+) will AgCl begin to precipitate?
(b) At what (Ag^+) will Ag_2CrO_4 begin to precipitate?
(c) Which salt will precipitate first?

SOLUTION: (a) AgCl ionizes: $AgCl = Ag^+ + Cl^-$

Its solubility product expression therefore is:

$$(Ag^+) \times (Cl^-) = 1.6 \times 10^{-10}$$

Let:

$$(Ag^+) = X.$$
$$(Cl^-) = 0.001 \text{ M}$$

Substituting into the solubility product expression, we have:

$$X \times 0.001 = 1.6 \times 10^{-10}$$
$$(Ag^+) = X = 1.6 \times 10^{-7} \text{ M.}$$

Thus, the (Ag^+) necessary to begin the precipitation of AgCl is 1.6×10^{-7} moles per liter.

(b) Ag_2CrO_4 ionizes: $Ag_2CrO_4 = 2\,Ag^+ + CrO_4^{--}$.

Its solubility product expression therefore is:

$$(Ag^+)^2 \times (CrO_4^{--}) = 9.0 \times 10^{-12}.$$

Let:

$$(Ag^+) = X$$
$$(CrO_4^{--}) = 0.005 \text{ M.}$$

Substituting into the solubility product expression:

$$X^2 \times 0.005 = 9.0 \times 10^{-12}$$
$$X^2 = \frac{9.0 \times 10^{-12}}{5.0 \times 10^{-3}} = 1.8 \times 10^{-9} = 18 \times 10^{-10}.$$
$$(Ag^+) = X = 4.2 \times 10^{-5} \text{ M}$$

Thus, the (Ag^+) necessary to begin the precipitation of Ag_2CrO_4 is 4.2×10^{-5} moles per liter.

(c) The results of parts a and b show that it takes less silver ion to cause AgCl to begin to precipitate as silver ion is added to this solution. Therefore AgCl will begin to precipitate first from the solution. Under actual conditions, the white salt, AgCl, will continue to precipitate as silver ion is added until the amount of chloride ion left in solution is negligible. By that time the concentration of silver ion will have increased in the solution so that the red salt, Ag_2CrO_4, will begin to precipitate. The first appearance of the red salt indicates that the chloride has all been measured by the silver. Thus, we have an important method of analyzing solutions for their chloride content.

One more application of equilibrium should be mentioned at this point. An electrolyte which is only slightly soluble in water may be put into solution if its ions can be made to form molecules of weak electrolytes. Consider the following situation. Calcium carbonate, $CaCO_3$, is only slightly soluble in water. It ionizes: $CaCO_3 = Ca^{++} + CO_3^{--}$. If an acid is added to the saturated $CaCO_3$ solution, the hydrogen ions from the acid will react with the carbonate ions to form molecules of carbonic acid, H_2CO_3. This upsets the equilibrium between the ions of $CaCO_3$ in solution and the ions of it in the solid crystals. The precipitation process (reaction to the left in the ionization equation) is interfered with, but the dissolving process (reaction to the right in the ionization equation) continues. The result is that all of the solid dissolves in the acidified solution. This phenomenon makes acid solutions good solvents for all salts except salts of strong acids.

Problem Set No. 11

1. 44.4 g. of $CaCl_2$ are dissolved in 200 g. of water. Assuming that the salt remains 100% ionized, find the temperature at which this solution will freeze.
2. What is the concentration of each of the ions present in each of the following solutions?

   (a) 0.05 M $MgCl_2$ (b) 0.25 M H_2SO_4
   (c) 0.1 M $Fe_2(SO_4)_3$

3. Which of the following pairs of electrolytes will react with each other in solution? Give a balanced equation for each reaction that takes place.

   (a) $K_2CO_3 + CaCl_2$ (c) $H_2SO_4 + AgNO_3$
   (b) $HNO_3 + Na_2SO_3$ (d) $HNO_3 + Na_2SO_4$

4. A 0.1 M solution of nitrous acid, HNO_2, is 6.3% ionized at room temperature. Find its ionization constant.
5. The ionization constant of acetic acid, $HC_2H_3O_2$,

is 1.8×10^{-5}. Find the (H^+) in a 0.1 M solution of this acid.

6. Find the percentage ionization of the solution in problem 5.

7. The ionization constant of HCN is 4×10^{-10}. Find the pH of a 0.1 M solution of HCN.

8. The ionization constant of NH_4OH is 1.8×10^{-5}. Find the pH of a 0.1 M solution of NH_4OH.

9. The ionization constant of acetic acid is 1.8×10^{-5}. To a 0.05 M solution of this acid, sufficient sodium acetate is added to raise the acetate ion concentration to 0.01 M. Find the pH of the resulting solution.

10. With the aid of ionization equations, show how the following salts will, or will not, hydrolyze, and indicate whether their solutions will be acidic, basic, or neutral.

    (a) KNO_3 (c) NH_4NO_3 (e) $FeCl_3$
    (b) KNO_2 (d) Na_2SO_3

11. The solubility of $Mg(OH)_2$, magnesium hydroxide, is 0.0009 grams per 100 ml. Find its solubility product constant.

12. The solubility product constant of silver acetate, $AgC_2H_3O_2$, is 2×10^{-3}. Find the concentration of each of the ions in a saturated solution of this salt.

13. The solubility product constant of magnesium carbonate, $MgCO_3$, is 1×10^{-5}. A solution has a carbonate ion concentration of 0.05 M. What concentration of magnesium ion will be required to begin the precipitation of $MgCO_3$?

14. Which of the following salts will dissolve in acid (use ionization equations to verify your answer)?

    (a) $MgCO_3$ (c) $Cu(OH)_2$
    (b) $AgCl$ (d) $BaSO_4$

15. The solubility product constant of calcium hydroxide, $Ca(OH)_2$, is 8×10^{-6}. Find the pH of a saturated solution of this substance.

SUMMARY

Electrolytes dissociate into **ions** in solution. **Acids** ionize to produce hydrogen or hydronium ions. **Alkalis** ionize to produce hydroxide ions.

Salts ionize, but produce neither of these two ions.

The chemistry of ions in solution is **independent of the source** of the ions.

A **strong electrolyte** is 100% ionized. A **weak electrolyte** is only partially ionized.

Ions in solution react if:

a) They can form a **weak electrolyte.**
b) They can form an **insoluble substance.**
c) They **oxidize or reduce** one another.

A **state of equilibrium** exists when two opposing processes take place simultaneously at the same rate.

The **Principle of Le Chatelier** is: A system in equilibrium, when subjected to a stress resulting from a change in temperature, pressure, or concentration, and causing the equilibrium to be upset, will adjust its position of equilibrium to relieve the stress, and reestablish equilibrium.

The **Law of Mass Action** is: the velocity of a reaction is proportional to the product of the molar concentrations of the reacting substances, taken to proper powers, the powers being the coefficients of the reactants in the balanced equation for the reaction.

The **pH of a solution** is the logarithm of the reciprocal of the molar concentration of the hydrogen ion. Neutral solutions have a pH of 7. Acidic solutions have a pH of less than 7. Basic solutions have a pH of greater than 7.

The **common ion effect** is: the addition of an ion in common with an ion of a solute represses the dissociation of the solute.

Hydrolysis is the reaction between the ions of an electrolyte and the ions of water in solution.

The **solubility product principle** is: At saturation, the product of the molar concentrations of the ions of the electrolyte, raised to proper powers, is a constant. The powers are the coefficients in the ionization equation of the electrolyte.

Substances only slightly soluble in water can be caused to dissolve if weak electrolytes can be formed in their saturated solution.

CHAPTER 10

OXIDATION-REDUCTION

Structurally, we may think of all chemical changes as falling into one of two categories:

1. Those which involve valence change, and
2. Those which do not.

If you look back over the reactions studied in the last chapter, you will observe that none of them involved valence change. The motivating force which caused each of those reactions to take place was the force of attraction which exists between oppositely charged ions. More specifically, the reactions occurred only when the forces of electrical attraction were strong enough to overcome the nullifying influence of the polar water molecules. The polarity of water was responsible for all dissociations. The high degree of electrical attraction between the ions of weak electrolytes and insoluble substances was responsible for the formation of these compounds. Let us now turn our attention to reactions which involve changes in valence numbers.

DEFINITIONS

Consider the following reaction:

$$2\,Cu + O_2 = 2\,CuO.$$

This reaction occurs when copper is heated in air, and the black oxide forms on the surface of the metal. Let us now write the valence number of each element under its symbol wherever it appears in the equation.

$$2\,Cu + O_2 - 2\,CuO.$$
$$\quad 0 \quad\ 0 \qquad +2,\ -2$$

As a result of the reaction, copper has changed valence from 0 to +2, and oxygen has changed valence from 0 to −2. A substance which has **gained in positive valence** has been **oxidized**. A substance which has **lost in positive valence** has been **reduced**. To state it in other terms:

Oxidation is a process involving an increase in valence number.

Reduction is a process involving a decrease in valence number.

Bearing in mind what we already know about the structure of matter and its relation to valence number, let us examine this reaction even more closely. How can an element show an increase in positive valence? The answer is quite simple. An element can increase its valence number only through the loss of electrons. Similarly, an element can show a decrease in valence number only through the gain of electrons. Therefore we can redefine our two processes in the following terms:

Oxidation is a process involving the loss of electrons.

Reduction is a process involving the gain of electrons.

In the reaction under consideration, the oxygen atoms gained electrons, and the copper atoms lost electrons. The substance that gains electrons in an oxidation-reduction reaction is known as the **oxidizing agent.** The substance that loses electrons in an oxidation-reduction is known as the **reducing agent.** Note that it was the free, elemental oxygen that acted as the oxidizing agent, and that free, elemental copper acted as the reducing agent in this reaction.

By this time you have probably observed that in oxidation-reduction reactions we are actually dealing with a **flow of electrons** from the reducing agent to the oxidizing agent. This is extremely important, for a flow of electrons is, by definition, **a current of electricity.** What causes such a current to flow, and some of the effects of such currents will be explored more fully in the next chapter.

One other facet of this reaction should be noted at this time. In any reaction involving valence change, both oxidation and reduction must be present. The one process provides the electrons which are absorbed by the other process. These reactions are often called **"Redox"** reactions, a term which is a shortened form of "oxidation-reduction.'' As a matter of fact, the Law of Conservation of Matter requires that **the total number of electrons gained in a reaction of this type must be equal to the total number of electrons lost.** Let us check this in the balanced reaction under consideration. Each copper atom loses 2 electrons, for a total of 4. Each oxygen atom gains 2 electrons, for a total of 4. Thus, the number of electrons gained and lost balances. This fact provides us with a powerful tool for balancing oxidation-reduction reactions.

BALANCING REDOX EQUATIONS

Let us look at the following skeleton equation and see how our principle of electronic balance can assist us in balancing the equation. The numbers above the symbols are the valence numbers of the elements in the skeleton equation. (Rules for finding valence numbers were discussed in Chapter 4.)

$$\overset{1,-1}{HCl} + \overset{4,-2}{MnO_2} = \overset{2,-1}{MnCl_2} + \overset{1,-2}{H_2O} + \overset{0}{Cl_2}$$

The changes in valence number for manganese and chlorine atoms are apparent from the valence numbers above the equation. The lines below the equation show that each manganese atom gains 2 electrons, while each chlorine atom loses 1 electron. In order for the number of electrons lost to be equal to the number of electrons gained, it will take 2 atoms of chlorine to react for each atom of manganese that reacts. So the equation tentatively becomes:

$$2\,HCl + MnO_2 = MnCl_2 + H_2O + Cl_2.$$

This balances the electrons, but we see that we need 2 more chlorine atoms for the $MnCl_2$. This will change the coefficient of HCl to 4, and of H_2O to 2:

$$4\,HCl + MnO_2 = MnCl_2 + 2\,H_2O + Cl_2.$$

Thus the equation is balanced.

Let us now look at a redox equation that involves ions.

$$\overset{2}{Fe^{++}} + \overset{7}{MnO_4^-} + H^+ = \overset{2}{Mn^{++}} + \overset{3}{Fe^{+++}} + H_2O.$$

The changes in valence number, and the gain and loss of electrons are indicated as in the previous example. In order for the gain and loss of electrons to balance, 5 ferrous ions will have to react for each permanganate ion that reacts. So we insert a coefficient of 5 in front of each of the iron ions, thus:

$$5\,Fe^{++} + MnO_4^- + H^+ = Mn^{++} + 5\,Fe^{+++} + H_2O.$$

This balances the electrons, so now we have to balance the hydrogen and oxygen atoms. We need 4 oxygen atoms on the right, which in turn requires 8 hydrogen atoms on the left. The equation then becomes:

$$5\,Fe^{++} + MnO_4^- + 8\,H^+ = Mn^{++} + 5\,Fe^{+++} + 4\,H_2O.$$

The equation now appears to be balanced. However, there is still one more point to be checked. In balanced ionic equations, the net charge on each side of the equation must be the same. Examining our equation, we see that a net charge of +17 appears on both sides of the equation. Therefore the equation is balanced.

Problem Set No. 12

1. Which of the following are oxidation-reduction reactions?

    (a) $2\,Fe^{++} + Cl_2 = 2\,Fe^{+++} + 2\,Cl^-$.
    (b) $Cu^{++} + H_2S = CuS + 2\,H^+$.
    (c) $2\,SO_2 + O_2 = 2\,SO_3$.

2. (a) Balance the following skeleton equation:

    $$H_2S + I_2 = S + I^- + H^+.$$

    (b) Which substance is the oxidizing agent?
    (c) Which substance is reduced?

3. (a) Balance the following skeleton equation:

$$Cu^{++} + I^- = I_2 + Cu^+.$$

(b) Which substance is the reducing agent?

(c) Which substance is oxidized?

(d) Why would the mere addition of a coefficient of 2 in front of I^- **not** balance this equation?

4. (a) Balance the following skeleton equation:

$$MnO_4^- + Sn^{++} + H^+ = Mn^{++} + Sn^{++++} + H_2O.$$

(b) Which substance is the oxidizing agent?

(c) Which substance is oxidized?

5. (a) Balance the following skeleton equation:

$$Cr_2O_7 + Fe^{++} + H^+ = Cr^{+++} + Fe^{+++} + H_2O.$$

(b) Which substance is the reducing agent?

(c) Which substance is reduced?

SUMMARY

Oxidation-reduction reactions involve a change in valence number.

Oxidation is a process involving an **increase in valence number** or a **loss in electrons**. **Reduction** is a process involving a **decrease in valence number** or a **gain in electrons**.

An **oxidizing agent gains electrons**. A **reducing agent loses electrons**.

A **flow of electrons** accompanies oxidation-reduction reactions.

The total number of electrons gained must equal the total number of electrons lost in an oxidation-reduction reaction.

An ionic redox reaction is balanced only when both the number of atoms and the net charge are equal on both sides of the equation.

CHAPTER 11

ELECTROCHEMISTRY

We have seen that matter has an electrical nature. The atoms that make up all matter consist of nuclei carrying positive electrical charges surrounded by electrons possessing negative electrical charges. Furthermore, we know that a relatively large class of substances, called electrolytes, consist essentially of charged particles called ions. Solutions of electrolytes are able to conduct an electric current. Finally, we learned in the last chapter that during all oxidation-reduction reactions a flow of electrons accompanies the reaction. These facts arouse one's curiosity as to what relationship there might be between electric currents and matter. Specifically, one might ask two questions:

1. What is the effect of an electric current on solutions of electrolytes?
2. What electrical effects accompany oxidation-reduction reactions?

Let us now explore each of these two areas.

ELECTRICAL UNITS

There are a few electrical units which we should understand before we proceed with our exploration. These are:

Coulomb. A coulomb is a unit **quantity** of electricity. It is that amount of electricity which will deposit 0.001118 grams of silver from a solution of silver nitrate.

Ampere. An ampere is a unit **rate of flow** of electricity. It is that current which will deposit 0.001118 grams of silver in one second. In other words, an ampere is a current of one coulomb per second.

Ohm. An ohm is a unit of electrical **resistance.** It is the resistance offered at 0° to an unvarying current by a column of mercury 106.3 cm. in length, of approximately 1 sq. mm. uniform cross-sectional area, and weighing 14.4521 grams.

Volt. A volt is a unit of **electromotive force.** It is the difference in electrical potential required to cause a current of one ampere to pass through a resistance of one ohm.

ELECTROLYSIS

Let us now turn our attention to the effects of an electric current on solutions of electrolytes. The passage of an electric current through a solution of an electrolyte is known as **electrolysis.** The effect of electrolysis is to decompose the electrolyte. Figure 29 shows a typical electrolysis setup. It consists of a **battery** whose terminals are each connected to an **electrode.** The electrodes are immersed in the solution of the **electrolyte.**

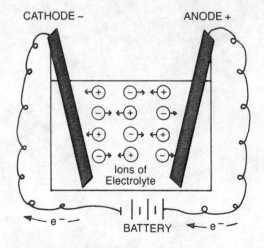

Fig. 29. Electrolysis Cell

In solution, electrolytes are dissociated into their ions. Positively charged ions are known as **cations.** Negatively charged ions are known as **anions.** The battery has the effect of pulling electrons out of one of the metallic electrodes and pushing them onto the other. This causes the first electrode, with the deficiency of electrons, to become positively charged. The other electrode, with the excess of electrons, becomes negatively charged. The positive electrode attracts the negative anions. This electrode is called the **anode.** The negative electrode attracts the positive cations. This electrode is called the **cathode.**

At the electrodes, chemical reactions take place. At the cathode, the positive cations absorb electrons from the negative cathode and become neutral atoms. At the anode, the negative anions yield their electrons to the positive anode and become neutral atoms. Thus, at both electrodes, ions are changed to free atoms. Since a change in charge, and consequently a change in valence number takes place in the reactions, all of these reactions are of the oxidation-reduction type. **Oxidation takes place at the anode. Reduction takes place at the cathode.**

Let us now consider a specific case of electrolysis. A solution of hydrochloric acid, HCl, will do to begin with. In solution, the HCl is ionized:

$$HCl = H^+ + Cl$$

In the electrolysis cell, the H^+ ions will migrate toward the cathode, and the Cl^- ions will migrate the anode. At the electrodes, the following reactions will take place:

(At cathode) $$H^+ + e^- = H$$

As you see, each hydrogen ion picks up an electron from the negatively charged cathode to become a hydrogen atom. Pairs of hydrogen atoms then unite to form molecules of hydrogen gas, H_2, which bubbles out of solution at the cathode.

(At anode) $$Cl^- = Cl + e^-$$

After the chloride ion loses its electron to the anode, pairs of chlorine atoms unite to form chlorine gas, Cl_2, which bubbles out of solution at the anode.

The net effect of this process is the decomposition of HCl into hydrogen and chlorine gases. The overall equation is:

$$2 HCl = H_2 + Cl_2.$$

Note that during electrolysis the solution remains electrically neutral. Molecules of chlorine and hydrogen gas are formed in equal number. The electrolysis will continue until the electrolyte has been completely decomposed.

One other feature of this electrolysis should be pointed out. Water molecules may also be oxidized at the anode as well as the chloride ions. This introduces another possible anode reaction into the system. It is:

$$2 H_2O = O_2 + 4 H^+ + 4 e^-.$$

This anode reaction requires more electrical energy than does the oxidation of the chloride ion, so when appreciable concentrations of chloride ions are present, the chloride ions will be **selectively oxidized** and this second reaction will not occur. However, near the end of the electrolysis, when the concentration of chloride is low, some oxygen will form to contaminate the chlorine gas.

Let us now consider the electrolysis of a solution of copper sulfate, $CuSO_4$. In solution, the $CuSO_4$ is ionized.

$$CuSO_4 = Cu^{++} + SO_4 .$$

The solvent, water, is also ionized. Its ions are also present in the solution. The following reactions are possible:

Cathode:	*Anode:*
$Cu^{++} + 2 e = Cu$	$SO_4 = 2 e + SO_4$
$H^+ + e = H$	$4 OH = 2 H_2O + O_2 + 4 e .$

The reaction which takes place at each electrode depends upon which reaction requires the **smaller amount of energy.** At the cathode, it takes less energy to reduce the copper ions than it does to reduce hydrogen ions. Therefore the copper will plate onto the cathode in this solution. Similarly, it takes less energy at the anode to oxidize the hydroxide ions than it does to oxidize the sulfate ions. Therefore oxygen gas will form at the anode. The overall reaction for this electrolysis may be written:

$$2 CuSO_4 + 2 H_2O = 2 Cu + O_2 + 2 H_2SO_4.$$

We will learn more about the nature of the **"preferential deposition"** of copper and oxygen observed in this example later in this chapter. It should likewise be pointed out that in both of the examples of electrolysis studied, the electrodes had to be made of a material, like platinum, that is inert to the substances in solution.

FARADAY'S LAWS

Michael Faraday, while studying the phenomenon of electrolysis, discovered that a definite quantitative relationship exists between the amounts of elements formed in electrolysis and the amount of electricity used in the process. His discoveries were formulated into two laws which bear his name. They are:

1. The weight of a given element liberated at an electrode during electrolysis is directly proportional to the quantity of electricity which passes through the solution.

2. When the same quantity of electricity passes through solutions of different electrolytes, the weights of the substances liberated at the electrodes are directly proportional to their equivalent weights.

The quantity of electricity required to deposit one equivalent weight of any element is 96,500 coulombs. This quantity of electricity is known as the **Faraday.** The number of coulombs in a Faraday can be found by dividing the equivalent weight of silver, 107.87, by the weight of silver deposited by one coulomb, 0.001118 grams. The following relationships are useful in solving problems relating to electrolysis.

$$\text{Number of coulombs} = \text{Number of amperes} \times \text{time in sec.} \quad (1)$$

$$\text{Number of Faradays} = \frac{\text{Number of coulombs}}{96,500} \quad (2)$$

Let us work through an example of a problem involving the application of Faraday's Laws.

EXAMPLE 1: A current of 20 amperes is passed through a solution of $CuSO_4$ for 2 hours. What weight of copper will be deposited?

SOLUTION: The atomic weight of copper is 63.55. The valence number of copper in $CuSO_4$ is +2. The equivalent weight of copper is $63.55/2 = 31.78$. The time in seconds is: $2 \times 60 \times 60 = 7200$ sec. The number of coulombs is: $20 \times 7200 = 144000$ coulombs. The number of Faradays is: $144000/96500 = 1.492$ Faradays. The weight of copper deposited is: $31.78 \times 1.492 = 47.42$ g.

NOTE: Each of these steps can be combined, and a single formula can thus be developed. It is:

$$\frac{\text{Wt. of}}{\text{deposit}} = \frac{\text{Eq. Wt.} \times \text{Amps.} \times \text{Time in sec.}}{96,500} \quad (3)$$

APPLICATIONS OF ELECTROLYSIS

The applications of electrolysis may be placed into three categories: analytical chemistry, production chemistry, and electroplating. The analytical methods based upon electrolysis utilize two different ideas connected with this phenomenon. As you know, electricity is carried through a solution of an electrolyte by the charged ions migrating to the electrodes. The more ions present, the better the conductivity. In many reactions of ions in solution, the total number of ions decreases as the reaction progresses. For example, if a solution of sulfuric acid is added to a solution of barium hydroxide, the products formed are water (a weak electrolyte) and barium sulfate (an insoluble substance). The reaction in ionic form is:

$$(2\,H^+.SO_4^{--}) + (Ba^{++}.2\,OH^-) = \underline{BaSO_4} + 2\,H_2O.$$

Thus when equivalent amounts of the two electrolytes have been reacted, the number of ions in solution reduces practically to zero. As the acid is added to the base, the conductivity of the solution diminishes, reaching a minimum at the equivalence point, and then increasing again as excess acid is added. From the volume of standard acid required to reach the point of minimum conductivity with a measured volume of unknown base, the concentration of the unknown base can be computed.

Other analytical methods are based on finding the weight of an element liberated by a known current in a measured amount of time. The equivalent weight of the unknown metal then is computed, which in turn leads to its identity.

Production chemistry probably represents the greatest economic contribution of the phenomenon of electrolysis. Many elements are produced commercially from electrolysis cells. Among these are hydrogen, oxygen, sodium, potassium, magnesium, calcium, aluminum, and many others. In addition, many important chemical compounds are obtained as by-products from these electrolyses. Either water solutions or **fused** (molten) baths of salts are used in the production of these substances by electrolysis. Many metals, such as copper, zinc, and silver, may be purified by electrolysis as the final step in their production.

In electroplating, the object to be plated is immersed in the electrolyte solution and is made the cathode of the cell. The electrolyte is a salt of the metal to be plated out, and the anode consists of a block of the metal to be plated out. The anode then goes into solution as the electroplating process proceeds. For example, in copper plating copper is used as the anode, and copper sulfate solution is the electrolyte. The possible anode reactions, then, are:

$$SO_4^{--} = 2\,e^- + SO_4$$
$$4\,OH^- = 4\,e^- + 2\,H_2O + O_2$$
$$Cu = 2\,e^- + Cu^{++}$$

The last reaction requires the least energy, and it therefore is the only one to occur. Copper ions are thus replaced in the solution as fast as they plate out on the object, and the concentration of the electrolyte remains constant. The weight of metal plating out is equal to the weight of metal that dissolves from the anode.

Table XXIV

Activity Series of Metals

Occurrence	Metal	Reactivity to		
		H_2O	HCl, dil. H_2SO_4	HNO_3, conc. H_2SO_4
Never found free in Nature	K-Potassium Ca-Calcium Na-Sodium	React with Cold Water To give H_2	Explosive action	Extremely explosive
	Mg-Magnesium Al-Aluminum Mn-Manganese Zn-Zinc Cr-Chromium	Much less active: Hot metal gives H_2 with steam	Liberate hydrogen gas	Liberate gas OTHER THAN HYDROGEN.
Rarely found free in Nature	Fe-Iron Co-Cobalt Ni-Nickel Sn-Tin Pb-Lead	Very poor activity with steam	Slow action Very slow action	NO, NO_2, or NH_3 from HNO_3. SO_2 or H_2S from conc. H_2SO_4.
	H-HYDROGEN			
Often found free in Nature	Sb-Antimony Bi-Bismuth Cu-Copper Hg-Mercury Ag-Silver	Inactive with water. No gas of any kind liberated.	Inactive with these acids	Action decreases as we go down list.
Found free in Nature	Pt-Platinum Au-Gold			Inactive.

ACTIVITY SERIES OF METALS

Let us now turn our attention to the electrical effects accompanying chemical change. Let us begin by examining the behavior of metals toward an acid like HCl. Some metals, when added to a water solution of this acid, will be dissolved by the acid and will liberate hydrogen gas from the solution. Others, like copper, silver, and gold, are not attacked by the acid, and do not liberate hydrogen from the solution. Furthermore, among the metals which do react with the acid solution, a great difference in rate of reaction exists. Sodium metal produces an explosion if dropped in the acid solution. Zinc causes a vigorous but quiet evolution of hydrogen gas. Lead liberates hydrogen only very slowly, at temperatures well above room temperature. Table XXIV is a list of metals in their order of reactivity, with the most active metal at the top. The table also summarizes other properties of these metals. It should be studied rather carefully.

Another important feature of the listing in Table XXIV is that **a metal higher on the list will replace the ions of metals lower on the list from solution, and liberate the lower metal in the free state.**

For example, if iron nails are dropped into a solution of copper sulfate, $CuSO_4$, the iron will be dissolved and the copper metal will precipitate from the solution. The reaction is:

$$Fe + Cu^{++} = Fe^{++} + \underline{Cu}.$$

EXPERIMENT 21: From your druggist obtain 1 oz. of copper sulfate crystals. Dissolve about 1/10 oz. of these crystals in a small glass of water. Thoroughly clean a nail with sandpaper. Place the nail in the copper sulfate solution. Let this stand overnight. You will observe the copper plate out on the surface of the nail and then fall to the bottom of the glass.

On the other hand, if a strip of copper metal is placed in a solution of ferrous sulfate, $FeSO_4$, **no reaction will take place.** Thus we can see that the activity series of metals has rather broad application in assisting us to predict whether or not a reaction of the type described will take place. It should be pointed out, by the way, that this **hydroprecipitation** of less active metals by more active ones is a process frequently used in extracting metals from their ores.

A glance at the reaction above quickly reveals that it is of an oxidation-reduction type. All reactions based on the activity series of metals are oxidation-reduction reactions. We now want to find out just what makes this type of reaction take place.

Let us analyze the following reaction:

$$Fe + Cu^{++} = Fe^{++} + \underline{Cu}.$$

We may think of this reaction as taking place in two steps:

$$Fe = Fe^{++} + 2\,e^-$$
$$Cu^{++} + 2\,e^- = \underline{Cu}.$$

The two parts may be called **half-reactions.** The first half-reaction shows iron metal going into solution by giving up two electrons and becoming ferrous ions. The second half reaction shows copper ions absorbing two electrons to become metallic copper atoms. Now note this point! As soon as the full reaction has proceeded to any extent at all, we have a system that contains **both metals and both ions** at the same time. Yet the reaction continues to proceed only one way, that is, the iron continues to dissolve and the copper continues to precipitate out. The only possible explanation for this is that **electrons continue to flow from the iron metal to the copper ions.** No electrons flow from the copper metal to either of the ions, nor do any electrons flow from the iron metal to the ferrous ions. Therefore **iron loses electrons more easily than copper,** and it loses them specifically to the copper ions.

In order for electrons to flow, thereby creating an electric current, an **electromotive force (emf)** must push the electrons along their path. A difference in electrical potential, and consequently an emf, exists between all oppositely charged bodies and between a charged body and a neutral one. Thus there is an emf between each metal and each of the ions in the system. However, since the electrons actually flow from the iron metal to the copper ions, the greatest emf must obviously be between these two particles.

The emfs associated with a large number of oxidation-reduction half-reactions have been

Table XXV

Half-Reaction (Electrode) Potentials

Half-Reaction	Emf (volts)	Half-Reaction	Emf (volts)
$Li = Li^+ + e^-$	+3.045	$4OH^- = O_2 + 2H_2O + 4e^-$	−0.401
$K = K^+ + e^-$	+2.924	$Ni(OH)_2 + 2OH^- =$	
$Ca = Ca^{++} + 2e^-$	+2.76	$\quad NiO_2 + 2H_2O + 2e^-$	−0.49
$Na = Na^+ + e^-$	+2.711	$2I^- = I_2 + 2e^-$	−0.535
$Mg = Mg^{++} + 2e^-$	+2.375	$2Mn_2O_9 + Zn(NH_3)_4^{++} + 2H_2O =$	
$Al = Al^{+++} + 3e^-$	+1.166	$\quad 4MnO_2 + 4NH_4^+ + Zn^{++} +$	
$Mn = Mn^{++} + 2e^-$	+1.185	$\quad 4e^-$	−0.738
$Fe + 2OH^- = Fe(OH)_2 + 2e^-$	+0.877	$Fe^{++} = Fe^{+++} + e^-$	−0.770
$H_2 + 2OH^- = 2H_2O + 2e^-$	+0.8277	$2Hg = Hg_2^{++} + 2e^-$	−0.7961
$Zn = Zn^{++} + 2e^-$	+0.7628	$Ag = Ag^+ + e^-$	−0.7996
$Cd + 2OH^- = Cd(OH)_2 + 2e^-$	+0.761	$2Br^- = Br_2 + 2e^-$	−1.065
$Cr = Cr^{+++} + 3e^-$	+0.74	$2H_2O = O_2 + 4H^+ + 4e^-$	−1.229
$Fe = Fe^{++} + 2e^-$	+0.4402	$2Cl^- = Cl_2 + 2e^-$	−1.3583
$Pb + SO_4^{--} = PbSO_4 + 2e^-$	+0.356	$2Cr^{+++} + 7H_2O =$	
$Co = Co^{++} + 2e^-$	+0.28	$\quad Cr_2O_7^{--} + 14H^+ + 6e^-$	−1.33
$Ni = Ni^{++} + 2e^-$	+0.23	$Au = Au^{+++} + 3e^-$	−1.42
$Sn = Sn^{++} + 2e^-$	+0.1364	$Mn^{++} + 4H_2O = MnO_4^- +$	
$Pb = Pb^{++} + 2e^-$	+0.1263	$\quad 8H^+ + 5e^-$	−1.491
$Fe = Fe^{+++} + 3e^-$	+0.036	$PbSO_4 + 2H_2O = PbO_2 +$	
$H_2 = 2H^+ + 2e^-$	+0.000	$\quad SO_4^{--} + 4H^+ + 2e^-$	−1.685
$Sn^{++} = Sn^{++++} + 2e^-$	−0.15	$SO_4^{--} = SO_4 + 2e^-$	−1.9
$Cu = Cu^{++} + 2e^-$	−0.345	$2F^- = F_2 + 2e^-$	−2.87

measured. Some of these are listed in Table XXV. Notice that in each of the half-reactions, the electrons are **given off.** In the table, the emf between hydrogen atoms and hydrogen ions is arbitrarily taken as **zero.** The substances above hydrogen in the list lose electrons more easily than hydrogen atoms do. They are thus given positive emfs. The substances below hydrogen lose electrons less readily than hydrogen atoms, so they are assigned negative emfs. Note further that the position of a substance in the activity series of metals corresponds with its position in Table XXV.

A full oxidation-reduction reaction is a combination of two half-reactions. In the full reaction, the half-reaction with the more positive emf will proceed from left to right as written in Table XXV. It is called the **reducing half-reaction,** and the substance losing the electrons is the reducing agent in the full reaction. The more negative half-reaction will turn around, or proceed from right to left as written in Table XXV. It is called the **oxidizing half-reaction,** and the substance absorbing the electrons is the oxidizing agent in the full reaction. Substances on the left side of the equations in Table XXV are reducing agents in the order of decreasing potency. Substances on the right sides of the equations in Table XXV are oxidizing agents in the order of increasing potency. Thus, elemental lithium is the most powerful reducing agent listed, and gaseous fluorine, F_2, is the most powerful oxidizing agent listed.

The total emf compelling the full reaction to take place will be the **algebraic difference** of the emfs of the two half-reactions involved. Let us see how this works out for the iron-copper system.

From Table XXV:

$$Fe = Fe^{++} + 2 e^- \qquad (+0.440 \text{ v.})$$
$$Cu = Cu^{++} + 2 e^- \qquad (-0.345 \text{ v.})$$

You will observe that the iron half-reaction has the more positive emf. Therefore it proceeds as written. Since the copper half-reaction has a more negative emf, it will proceed from right to left as written, or will turn around. The full reaction may then be found as follows:

$$Fe = Fe^{++} + 2 e^- \quad \text{(Reducing half-reaction)}$$
$$Cu^{++} + 2 e^- = Cu \quad \text{(Oxidizing half-reaction)}$$

Adding: $Fe + Cu^{++} = Fe^{++} + Cu.$ (Full reaction)

The electrons cancel out during the addition process. The total emf for this reaction will be:

$$0.440 - (-0.345) = 0.785 \text{ volts.}$$

The magnitude of the total emf gives an indication of the vigor with which the reaction will proceed. Substances close together in Table XXV will be impelled to react by only a small total emf, and consequently will react only slowly and mildly. Substances far apart in Table XXV will be impelled to react by a larger total emf, and will react much more rapidly and vigorously.

VOLTAIC CELLS

The flow of electrons in oxidation-reduction reactions presents an interesting possibility. If the flow can be harnessed in some way and caused to pass through wires, we would have a means of obtaining electrical energy from chemical action. Figure 30 illustrates one method of harnessing the electricity from a chemical change. The arrangement is known as a **voltaic cell,** and consists essentially of solutions of two different ions.

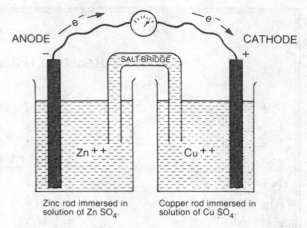

Fig. 30. Voltaic Cell

Strips of metal of the same substance as the positive ions of the solution are immersed into each solution. A salt bridge, which is a solution of a third electrolyte which does not react with either of the two original solutions, is placed between the two solutions to enable ions to migrate from one solution to the other. A wire would not do as a replacement for the bridge, for it would not permit the passage of ions. When the two strips of metal are connected by an external circuit, a current flows through the circuit. The emf of the cell depends upon the substances used in it.

In Figure 30, the two half-reactions involved are:

$$Zn = Zn^{++} + 2 e \qquad (+0.762 \text{ v.})$$
$$Cu = Cu^{++} + 2 e \qquad (-0.345 \text{ v.})$$

Since the zinc half-reaction is more positive, it proceeds as written. Zinc atoms in the zinc rod go into solution as ions, and the electrons thus given up pass from the zinc rod, through the external circuit, over to the copper rod. Here they attract copper ions which absorb them and plate out as atoms on the rod. Thus the second half-reaction is turned around. The total reaction is:

$$Zn = Zn^{++} + 2 e^-$$
$$Cu^{++} + 2 e^- = Cu$$
$$\overline{Zn + Cu^{++} = Zn^{++} + \underline{Cu}.}$$

The total emf is:

$$0.762 - (-0.345) = 1.107 \text{ volts.}$$

Many other combinations of half-reactions from

Table XXV can be arranged into a voltaic cell of the type illustrated in Figure 30.

DRY CELLS

The common **dry cell** is another arrangement from which electrical energy is obtained from chemical action. Figure 31 shows a cross section of a dry cell. Notice that water is an essential ingredient in the "dry" cell. If water were not present in the paste, the ions could not migrate and the cell would be ineffective.

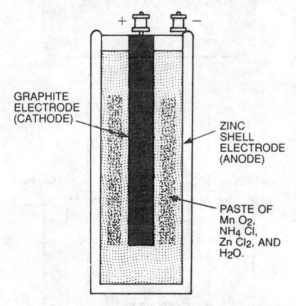

GRAPHITE ELECTRODE (CATHODE)

ZINC SHELL ELECTRODE (ANODE)

PASTE OF Mn O_2, NH_4 Cl, Zn Cl_2, AND H_2O.

Fig. 31. Dry Cell

The half-reactions in the dry cell may be written:

$$Zn = Zn^{++} + 2 e^- \qquad (+0.762 \text{ v.})$$
$$2 Mn_2O_3 + Zn(NH_3)_4{}^{++} + 2 H_2O = 4 MnO_2 +$$
$$4 NH_4{}^+ + Zn^{++} + 4 e^- \quad (-0.738 \text{ v.})$$

The total emf of the dry cell is:

$$0.762 - (-0.738) = 1.5 \text{ volts.}$$

The second half-reaction, which proceeds from right to left, actually takes place in steps as follows:

Step 1. The electrons from the zinc shell are absorbed by the ammonium ions in the paste:

$$2 NH_4{}^+ + 2 e^- = 2 NH_3 + H_2.$$

Step 2. The hydrogen gas thus formed tends to accumulate around the graphite electrode. Since the electrode potential of hydrogen is intermediate between the emfs of the two half-reactions of the dry cell (see Table XXV), it tends to set up its own voltaic action within the cell. This greatly reduces the voltage of the dry cell. The phenomenon is known as **polarization.** To offset this tendency, manganese dioxide, MnO_2, is included in the paste. This oxidizes the hydrogen gas as it is formed, thus acting as a **depolarizer:**

$$2 MnO_2 + H_2 = Mn_2O_3 + H_2O.$$

Step 3. The ammonia gas formed in step (1) reacts with the zinc ions from the zinc shell:

$$Zn^{++} + 4 NH_3 = Zn(NH_3)_4{}^{++}.$$

This third step has the effect of keeping the concentration of zinc ions in the paste constant and equal to the amount originally added to the paste in the form of zinc chloride, $ZnCl_2$.

It should be pointed out that the depolarization produced by the MnO_2 takes place rather slowly. Consequently if a dry cell is used to produce a great amount of electric current in a short time, hydrogen gas will accumulate faster than the MnO_2 can take care of it. Consequently the cell will polarize and its voltage will diminish. However, if the cell is permitted to sit idle for a period of time, it will recover its emf as a result of the continuing action of the MnO_2 in oxidizing the hydrogen gas.

LEAD STORAGE CELLS

Still another extremely important arrangement for producing an electric current from chemical action is to be found in the **lead storage cell.** This cell consists essentially of two lead gratings, one impregnated with spongy lead to provide abundant surface for reaction, and the other impregnated with lead dioxide, PbO_2, which serves as the second electrode. These gratings are then immersed in a solution of sulfuric acid, H_2SO_4. The half-reactions of the lead storage cell are:

$$Pb + SO_4^{--} = PbSO_4 + 2\,e^-$$
$$(+0.355\ v.)$$
$$PbSO_4 + 2\,H_2O = PbO_2 + SO_4^{--} + 4\,H^+ + 2\,e^-$$
$$(-1.685\ v.)$$

The total emf of the cell is:

$$0.355 - (-1.685) = 2.04\ \text{volts}.$$

An automobile battery has six of these cells in series producing 12 volts.

The outstanding feature of the lead storage cell is that it can be recharged. The two half-reactions are reversible. During the **discharge** of electricity from the cell, the two half-reactions proceed as follows:

(At anode)
$$Pb + SO_4 = PbSO_4 + 2\,e$$
(At cathode)
$$PbO_2 + SO_4 + 4\,H^+ + 2\,e = PbSO_4 + 2\,H_2O$$
(total)
$$Pb + PbO_2 + 4\,H^+ + 2\,SO_4 = 2\,PbSO_4 + 2\,H_2O.$$

Note that both half-reactions produce lead sulfate, $PbSO_4$, which deposits as crystals on both of the gratings in the cell. During the **recharge** process the lead storage cell acts as an electrolysis cell, for a source of electrical energy with an emf greater than the voltaic emf of the cell is attached to the cell and electrical energy is forced back into the cell. Each of the half-reactions above proceeds from right to left under the influence of this external emf. The $PbSO_4$ crystals are dissolved and the Pb, PbO_2, and H_2SO_4 are regenerated in the cell. (See Figure 32.)

Another interesting feature of the lead storage cell is the fact that its state of charge can readily be determined. In the discharge process H_2SO_4, the electrolyte in the cell, is used up. During the charging process it is regenerated. Now sulfuric acid is a very dense liquid, and the specific gravity of the solution of electrolyte changes appreciably as the cell is in use. At full charge, the specific gravity of the electrolyte is 1.28. When the cell is discharged, the specific gravity may drop to as low as 1.15. A **hydrometer,** which measures specific gravity of liquids, may then be used to indicate the extent to which a cell has been discharged.

The crystals of lead sulfate may cause a lead storage cell to "go dead" if they grow too large

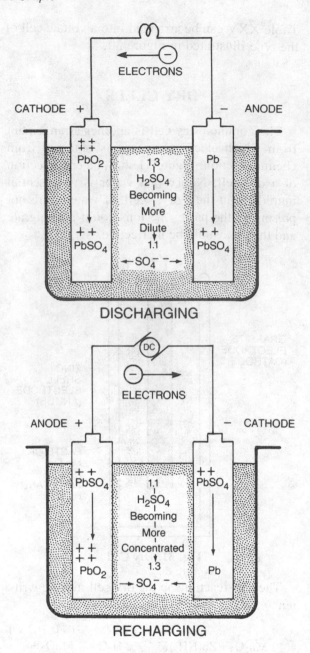

DISCHARGING

RECHARGING

Fig. 32. The Storage Battery

or cover the entire electrode. Factors promoting the growth and deposition of these crystals are:

1. Excessive heat applied to the cell.
2. Excessive loss of water from the solution of electrolyte.
3. Excessive and rapid "drain" on the cell caused when too much electrical energy is withdrawn from it.

Scientists are in general agreement that so-called battery additives, which usually consist of mixtures of sodium and magnesium sulfates (Glauber's salt and Epsom salt), are ineffective in prolonging the life of a lead storage cell.

When a lead storage battery is charged, water is electrolyzed from the sulfuric acid solution. From time to time this water must be replenished. Modern, **maintenance-free lead storage batteries** are designed so that water does not have to be added. They are made so that the negative plate has more capacity than the positive plate, thus oxygen is released from the positive plate before hydrogen is formed on the negative plate. The oxygen travels to the spongy lead negative plate where the reactions occur:

$$2\ Pb + O_2 = 2\ PbO$$
$$PbO + H_2SO_4 = PbSO_4 + H_2O.$$

Thus the water is reformed and remains in the battery.

NICKEL-CADMIUM CELLS

In the **nickel-cadmium cell,** the anode is cadmium metal. The cathode contains nickel oxide, NiO_2. The half-reactions in a solution of potassium hydroxide are:

(At anode)
$$Cd + 2\ OH^- = Cd(OH)_2 + 2\ e^- \qquad (+0.761\ v.)$$
(At cathode)
$$NiO_2 + 2\ H_2O + 2\ e^- = Ni(OH)_2 + 2\ OH^-\ (+0.49\ v.)$$

(total)
$$Cd + NiO_2 + 2\ H_2O = Cd(OH)_2 + Ni(OH)_2\ \ (1.251\ v.)$$

The $Cd(OH)_2$ and $Ni(OH)_2$ are formed on the Cd and NiO_2 electrodes. Therefore, unlike dry cells, nickel-cadmium cells can be recharged, and can be used repeatedly. In addition, unlike lead storage batteries, they can be made in a very small size, so that they are especially useful for small, portable electronic equipment such as hand calculators.

CORROSION

Not all of the effects of voltaic action are useful. Tiny voltaic cells which form in metals are responsible for the corrosion of these metals, causing an economic loss amounting to millions of dollars each year. All commercial iron and steel, for example, contains impurities, principally carbon. These impurities are not uniformly distributed in the metal, but are segregated at various points. As with all chemical substances, a difference in electrical potential exists between the atoms of the metal and the atoms of the impurities. When the metal is in contact with moist air, a film of water forms on the surface of the metal. Carbon dioxide in the air dissolves in this moisture and forms a solution of an electrolyte, H_2CO_3, through which ions can migrate. Since iron is more electropositive than its impurities, iron atoms act as the anode in the cell. They become oxidized and go into solution as ions. The liberated electrons pass through the metal to the atoms of the impurity, which acts as a cathode. Hydrogen ions from the water migrate to the cathode, pick up an electron to form hydrogen atoms, and are oxidized by atmospheric oxygen to water. This prevents polarization of the cell and permits the action to continue. The iron ions are likewise oxidized further by atmospheric oxygen, and then react with the water to form a complex hydrated oxide, $Fe_2O_3 \cdot X(H_2O)$, known as **rust.** Perfectly dry iron does not rust because voltaic action does not take place in the absence of water.

EXPERIMENT 22: Thoroughly clean four nails with sandpaper. Then:

(a) Place the first in a glass of tap water so that the nail is completely covered with water. Then cover the glass to prevent excessive evaporation.
(b) Place the second nail in a glass of water so that the nail is only partly covered with water. Again cover the glass to prevent excessive evaporation.
(c) Boil some water for 5 minutes in a Pyrex dish, add the third nail, and continue to boil for a few more minutes. Then remove the flame and quickly pour molten paraffin or vaseline over the water to exclude air.
(d) Dissolve a teaspoon of lye in a cup of water. Put the solution in a soda bottle and cork the bottle. Shake the solution in the corked bottle thoroughly and then permit it to stand overnight. Boil a nail in water exactly as in (c). Force the nail through the cork so that part of it is

exposed inside the bottle and part of it is exposed to the air above the cork.

Permit all four nails to stand a few days. Then you will observe the following:

Rusting in (a) where we have much water and little air (dissolved in the water).

Rusting in (b) where we have little water and much air.

No rusting in (c) where we have no air.

No rusting inside the bottle in (d) where we have no CO_2, but rusting outside the bottle.

EXPERIMENT 23: From your druggist obtain a few crystals of potassium ferricyanide. Heat one cup of water to boiling, add $1\frac{1}{2}$ teaspoons of clear gelatin, $\frac{1}{8}$ teaspoon of table salt, $\frac{1}{8}$ teaspoon of potassium ferricyanide crystals, and 8–10 drops of phenolphthalein indicator (prepared in Experiment 5). Pour this solution over a thoroughly cleaned nail in a Pyrex dish. Let stand overnight. You will observe the nail being corroded. Where the iron is going into solution, deep blue ferrous ferricyanide will form, indicating the anodes. Other parts of the nail will become cathodic and will cause hydroxide ions to concentrate around the cathodes. This will cause the phenolphthalein to turn pink in these areas.

The combatting of the corrosion of iron, and metals in general, is a huge enterprise today. A number of methods are used, all designed to prevent or overcome localized voltaic action in the metal.

1. Much progress has been made in producing and fabricating more homogeneous metals. When impurities are highly segregated, the cells can work in series and build up appreciable currents. Uniform distribution of impurities tends to nullify this collective effect and to minimize the effectiveness of the tiny voltaic cells.

2. Another method involves coating the metal with a film which will prevent contact between the metal and moisture in the air, thus preventing voltaic action from taking place. Various types of anticorrosion paints are used, and they are effective as long as the film remains intact. However, if the metal is subjected to temperature changes it will expand and contract to a greater extent than the coating, especially if the paint is thoroughly dried out. Cracks develop in the coating, and voltaic action is able to begin. New paints

are continually being developed to provide more adequate surface protection.

3. Another process is known as **galvanizing.** In this process iron is coated with a layer of zinc. Zinc is above iron in the activity series, but it oxidizes only superficially in the atmosphere. A thin film of zinc oxide, ZnO, forms on its surface. This film is so cohesive that it cannot be further penetrated by oxygen. Thus the zinc protects the iron. The protection continues even when the zinc coating is broken. Moisture and carbon dioxide penetrate any crack that develops in the coating. They form a voltaic cell with the zinc and iron. Since zinc is more electropositive, it passes into solution as zinc ions. These react with the hydroxide ions of water to form zinc hydroxide, $Zn(OH)_2$, which in turn combines with the dissolved carbon dioxide to form a basic zinc carbonate, $Zn_2(OH)_2CO_3$. This compound, which is very insoluble, forms a tight film similar to the ZnO film. It too is impervious to water and the atmospheric gases. Thus the crack is plugged and the iron remains protected.

4. A similar method involves coating the iron with tin. However, **tin-plating** functions differently from galvanizing. Tin, like zinc, oxidizes only superficially in the atmosphere, and as long as the coat remains intact, the iron is adequately protected. But if a crack develops in the tin plate, moisture and carbon dioxide enter the crack and form a voltaic cell with tin and iron. Since iron is higher in the activity series, it passes into solution by voltaic action. The tin then serves to accelerate the corrosion of the iron. Tin plate is used in food containers rather than the more effective zinc because zinc may react with the food to produce poisonous compounds.

5. Iron and steel may also be protected from corrosion by a method which uses **"sacrificial anodes."** In this method, metals such as zinc or magnesium, which are more electropositive than iron, are placed in the vicinity of the iron to be protected and are wired to it. The active metal will exhibit a greater emf toward oxidizing agents than the iron. Consequently they will be corroded and the iron will remain protected as long as any of the sacrificial metal remains. This method is particularly effective in protecting underground pipelines and underwater fittings, like

propellers, of ships. In voltaic cells, one metal is known as the **active** metal and the other is often called the **noble** metal. Yet, we see that the active metal is sacrificed to protect the noble one.

The principle of selective corrosion of sacrificial anodes is similar to the situation we encountered earlier in this chapter in studying electrolysis. We saw that when more than one electrode reaction was possible, one reaction preferentially occurred. All electrode reactions are half-reactions. In all cases, the half-reaction **with the more positive emf** will take place. Anode reactions always involve the loss of electrons. Therefore, at the anode, the half-reaction listed highest in Table XXV will occur. Cathode reactions involve the gain of the electrons. Therefore, since cathode reactions proceed from right to left in Table XXV, the sign of the emf of the half-reaction is reversed, and the half-reaction listed **lowest** in Table XXV will take place.

One final point should be mentioned. Matter has an electrical structure. The various forms of matter are impelled to react by electrical phenomena. Two different types of forces drive chemical substances into reaction. They are:

Force of electrostatic attraction. This force is responsible for all ionic reactions of a non-oxidation-reduction type.

Electromotive force. This force produces a current of electricity which oxidizes one substance and reduces the other.

Problem Set No. 13

1. In the electrolysis of a solution of sodium sulfate, Na_2SO_4, the following electrode reactions are possible:

At Anode: $SO_4^{--} = SO_4 + 2\ e^-$ $(-1.9\ \ \text{v.})$
$2\ H_2O - O_2 + 4\ H^+ + 4\ e^-$ $(-1.229\ \text{v.})$

At cathode: $Na^+ + e^- = Na$ $(-2.711\ \text{v.})$
$2\ H_2O + 2\ e^- = H_2 + 2\ OH^-$ $(-0.828\ \text{v.})$

  (a) Which electrode reactions will occur in each case?

  (b) What is the net reaction for the cell?

2. In the electrolysis of fused NaCl, the electrode reactions are:

At Anode: $2\ Cl^- = Cl_2 + 2\ e^-$ $(-1.358\ \text{v.})$

At cathode: $Na^+ + e^- = Na$ $(-2.711\ \text{v.})$

  (a) What is the minimum voltage required of a battery to cause the electrolysis of this cell?

  (b) Would a single dry cell produce sufficient emf to cause this electrolysis?

  (c) Would an automobile battery produce sufficient voltage to cause this electrolysis?

3. A current of 30 amperes is passed through a bath of fused calcium chloride for 1 hour. What weight of metallic calcium will be deposited on the cathode?

4. How long will it take a current of 20 amps to deposit 40 grams of metallic sodium from fused NaCl?

5. What volume of Cl_2 at standard conditions will be liberated by a current of 15 amperes passing through fused NaCl for 1 hour?

6. Which of the following reactions will take place? Write balanced equations for those which do take place.

  (a) $Mg + NiCl_2$ (d) $Ag + HCl$

  (b) $H_2 + AuCl_3$ (e) $Al + CuSO_4$

  (c) $Cu + ZnCl_2$ (f) $Cu + AgNO_3$

7. A voltaic cell consists of aluminum metal in $Al(NO_3)_3$ solution joined to lead metal in $Pb(NO_3)_2$ solution.

  (a) What two half-reactions are involved?

  (b) What is the total emf of the cell?

  (c) Which is the oxidizing half-reaction?

8. A voltaic cell consists of lead metal in $Pb(NO_3)_2$ solution joined to silver metal in $AgNO_3$ solution.

  (a) What two half-reactions are involved?

  (b) What is the total emf of the cell?

  (c) Which is the reducing half-reaction?

9. In the Edison storage cell, the electrodes consist essentially of iron and nickel dioxide. The half-reactions are:

$Fe + 2\ OH^- = Fe(OH)_2 + 2\ e^-$ $(+0.877\ \text{v.})$
$Ni(OH)_2 + 2\ OH^- = NiO_2 + 2\ H_2O + 2\ e^-$
                         $(-0.49\ \text{v.})$

The electrolyte in this cell is a solution of potassium hydroxide, KOH.

  (a) Which electrode is the anode?

  (b) What is the total emf of the cell?

10. For the balanced oxidization-reduction reaction:

$$Cr_2O_7^{--} + 6\,Fe^{++} + 14\,H^+ = 2\,Cr^{+++}$$
$$+ 6\,Fe^{+++} + 7\,H_2O.$$

(a) Write the half-reactions involved with the aid of Table XXV.
(b) What is the total emf impelling this reaction?

SUMMARY

Electrolysis is the utilization of electrical energy to obtain chemical action. When an electric current is passed through a solution of an electrolyte, the electrolyte is decomposed at the electrodes in an oxidation-reduction type reaction.

When more than one reaction is possible at an electrode, the one which requires the **least** amount of electrical energy will take place.

Faraday's Laws reveal that the amount of a substance produced by the electrolysis of a substance depends upon the **quantity of electricity** used and upon the **equivalent weight** of the substance.

The **Activity Series of Metals** lists the metals in the order of their chemical reactivity in oxidation-reduction reactions. A metal higher on the list will replace the ions of the metals lower on the list from solution, and will liberate the lower metal in the free state.

Oxidation-reduction reactions consist of two **half-reactions,** one giving off electrons and the other absorbing them. An **electromotive force** accompanies each half-reaction. The emf of the total reaction is the **algebraic sum** of the emfs of the two half-reactions.

A **voltaic cell** converts chemical energy into electrical energy. Several types of voltaic cells, such as **dry cells, nickel-cadmium** and the **lead storage cell,** produce electrical energy in useful form.

The **corrosion of metals** is caused basically by the voltaic action of tiny cells that develop on the surface of the metal. The metal is oxidized by this action. The methods of combatting corrosion involve the elimination of voltaic action from the surface of the metal.

Matter, being of electrical nature, is impelled to undergo chemical change by either of two electrical phenomena:

a) **Forces of electrostatic attraction.**
b) **Electromotive forces.**

CHAPTER 12

THE ATMOSPHERE

Surrounding the earth is a sea of gas known as the **atmosphere** or **air.** The individual gases contained in the atmosphere are invisible, but we can feel their presence as we swing our hand through air or breathe it deeply. The principal substances present in the atmosphere are nitrogen, oxygen, carbon dioxide, and water vapor. Except for water vapor, which varies considerably, the atmosphere has a remarkably constant composition. Table XXVI summarizes the composition of the atmosphere exclusive of water vapor.

Table XXVI
Composition of Dry Air

Substance	Percentage by Volume Sea Level
Nitrogen	78.08
Oxygen	20.95
Argon	0.93
Carbon dioxide	0.03
Neon	0.0018
Helium	0.0005
Methane	0.0002
Krypton	0.00011
Hydrogen	0.00005
Nitrous oxide	0.00005
Xenon	0.000009
Ozone	Mere trace
Radon	Mere trace
Pressure	760 torr

From 25 to 50 miles up, the composition of the air remains about the same, but the air is much thinner and contains more ozone, particularly above 45 miles.

EXPERIMENT 24: Cut a cork in half the long way. Permit wax from a birthday candle to fall onto the flat cut surface of the cork and then stand a ½ inch length of small birthday candle in the wax. This is a boat. Float this in a small pot of water, light the candle, and then cover the entire boat with a large glass so that the open end of the glass extends well down into the water. You will observe that the water will immediately rise inside the inverted glass, and that the candle will become extinguished when about ⅕ the air in the glass is replaced with water.

In addition to water vapor, other variable components of the atmosphere are dust, pollen and other biological organisms, ammonia, oxides of nitrogen and sulfur, hydrogen sulfide, ozone, hydrocarbons, and miscellaneous substances resulting from geologic, vegetative, and industrial processes.

The atmosphere is divided into three layers. That part of it in contact with the earth and extending upward about 6 miles is known as the **troposphere.** About half the total weight of the atmosphere is contained in the troposphere. Moreover, all the water vapor in the atmosphere is to be found here. Since water vapor and weather are intimately related, all weather phenomena take place in the troposphere. From about 6 miles up to about 35 miles up is a layer known as the **stratosphere.** Aircraft on long flights normally use the lower portion of the stratosphere because it is free from clouds, storms, lightning, thunder, and all other forms of weather. Most meteorites are burned up in the stratosphere. In the lower stratosphere, about 12 to 15 miles up, ozone attains its maximum concentration in the atmosphere. This gas absorbs much harmful solar radiation. Above the stratosphere is the extremely rarefied region known as the **ionosphere.** This is the region in which the **aurora borealis,** or northern lights, stages its show. In the ionosphere, matter is so energized by solar radiation that it becomes ionized. Several ionic layers appear to exist, and they have the property of being able to reflect radio waves back to earth, thus making possible long-range transmission. The ionosphere may extend upward to 200 miles above the earth.

Air is a mixture despite its relatively constant composition. Each component of air retains its own unique physical and chemical properties, and air can be separated into its components by physical means. Normally differences in boiling points of the components are utilized in separating the substances in air.

The physical properties of the constituents of air are listed in Table XXVII, except those of the noble gases, which have already been presented in Table IX on p. 23. Note that air has a sufficiently constant composition to have some physical properties of its own. The molecular weight listed for air is the weight in grams of 22.4 liters of air at standard conditions.

The amount of water vapor in the atmosphere may vary from a mere trace on a cold dry day to well over 7% of the composition of the atmosphere on a hot, humid day. Normally the amount of water vapor in air is specified as the **relative humidity,** which is the ratio of the partial pressure of water vapor in air to the vapor pressure of water at the temperature of the air expressed as a percent. For example, if the partial pressure of water vapor in air at 77° F. (25° C.) is 16.5 torr, we can use Table XIV on p. 43 to compute the relative humidity of the air. In Table XIV we find that the saturation vapor pressure of water at 25° C. is 23.8 torr. The relative humidity therefore is:

$$\frac{16.5}{23.8} \times 100 = 69.3\%.$$

Notice that in the example just worked out, if the temperature of the air were to drop from 25° C. to 19° C., the air would then be saturated with water vapor. The temperature to which a given mass of air must be lowered to saturate it with water vapor is known as the **dew point.** The dew point of air is rather easy to measure. If a wet piece of cloth is attached to the bulb of a thermometer, the cooling caused by the evaporation of moisture from the cloth will lower the temperature reading to the value of the dew point. A second thermometer can then be used to measure the actual temperature of the air. With these two temperatures and the aid of Table XIV, the relative humidity can quickly be calculated. The vapor pressure associated with the dew point is the partial pressure of water vapor in the atmosphere. The vapor pressure associated with the actual air temperature is the saturation vapor pressure. The relative humidity is then computed from the following relationship:

$$\frac{\text{Partial pressure of water vapor}}{\text{Saturation vapor pressure}} \times 100 = \text{relative humidity} \quad (1)$$

Table XXVII
Physical Properties of Components of Air

Substance	Formula	Molecular Weight	Melting Point, °C.	Boiling Point, °C.	Solubility in Water, 0 °C. cc/100 ml	Density, 0 °C. g/liter
Nitrogen	N_2	28	−209.86	−195.8	2.33	1.251
Oxygen	O_2	32	−218.4	−183.0	4.89	1.429
Carbon dioxide	CO_2	44	− 56.6*	− 78.5†	171.3	1.977
Water vapor	H_2O	18	—	—	—	0.804
Hydrogen	H_2	2	−259.14	−252.8	2.14	0.090
Ozone	O_3	48	−192.5	−112	49	2.144
Air	—	29	—	—	—	1.293

* At a pressure of 5.2 atm.
† Sublimes (passes directly from solid to gas)

The relative humidity of air is an important factor in human comfort. The expression, "It's not the heat, it's the humidity" is based on fact. If the relative humidity is below 50%, air temperatures as high as 80° F. will feel cool. The coolness, of course, is the result of evaporation of moisture from the skin. But air becomes "close" or stuffy and highly uncomfortable when the relative humidity hovers near the 90% mark. We have learned to control indoor climate by a process known as **air conditioning.** The steps in the usual air conditioning process are as follows:

1. **Dehumidification** of the air by passing it over a desiccant such as **silica gel** or **calcium chloride.**
2. **Chilling** the air to the temperature of the desired dew point.
3. **Saturation** of the air with water vapor at this low temperature by bubbling it through water.
4. **Warming** the air back to the desired room temperature.

The amount of moisture in the air can also be controlled by a somewhat different process, based upon the properties of saturated solutions. Air of the desired room temperature is bubbled through a saturated solution of a salt which is in equilibrium with excess solute. As you know, a saturated solution has a definite vapor pressure at a given temperature. If the partial pressure of the moisture in the air is **greater** than the vapor pressure of the saturated solution, the solution will **absorb** moisture from the air. But if the partial pressure of the moisture in the air is **less** than the vapor pressure of the saturated solution, the solution will.**add** moisture to the air. Thus, by selecting the proper salt solution, **exact humidity control** can be maintained.

Let us now consider in more detail each of the components of the atmosphere.

NITROGEN

OCCURRENCE: The atmosphere is the only important source of free nitrogen. However, since the atmosphere consists of about 78% nitrogen, the supply is quite abundant. The principal source of combined nitrogen is the **guano** deposits found along the coast of Chile. These bird droppings which have accumulated in large quantity over centuries contain as much as 50% sodium nitrate, $NaNO_3$.

PREPARATION: **Commercially,** practically all nitrogen is obtained from the fractional distillation of liquid air. Since nitrogen has a lower boiling point than oxygen, it boils away from liquid air first and is then collected and compressed into tanks. This nitrogen is contaminated by the noble gases present in trace quantities in air, but is sufficiently pure for most uses.

In the laboratory, nitrogen is obtained by heating a mixture of sodium nitrite, $NaNO_2$, and ammonium chloride, NH_4Cl, in solution. The essential reaction is:

$$NH_4^+ + NO_2^- = N_2 + 2 H_2O.$$

PHYSICAL PROPERTIES: These are summarized in Table XXVII.

CHEMICAL PROPERTIES: The outstanding characteristic of nitrogen is its **high stability** and **relative inertness.** It combines with other elements only with difficulty. The process of inducing nitrogen to combine chemically with other substances is known as the **fixation** of nitrogen. This is accomplished in the following ways:

1. The very active metals will combine directly with nitrogen at high temperatures to form **nitrides:**

$$3 Mg + N_2 = Mg_3N_2.$$

2. In the **cyanamide** process, nitrogen and hot calcium carbide react to give calcium cyanamide and carbon:

$$CaC_2 + N_2 = CaCN_2 + C.$$

The hot cyanamide then reacts with steam under pressure to form ammonia:

$$CaCN_2 + 3 H_2O = CaCO_3 + 2 NH_3.$$

3. In the **Haber process,** nitrogen and hydrogen combine directly in the presence of a finely divided metallic catalyst at high temperature (about 500° C.) and extreme pressure (about 500 atm.):

$$N_2 + 3 H_2 = 2 NH_3.$$

4. The **nitrogen cycle** in nature provides the essential nitrogen compounds for plant life. **Bacteria** in the roots of leguminous plants like clover and peas are able to oxidize nitrogen into proteins, which in turn decompose to form nitrates in the soil. Other bacteria cause the decay of plant and animal tissue producing ammonia and free nitrogen which returns to the atmosphere. The ammonia is oxidized to nitrates by still other bacteria. In addition, much nitrogen combines directly with oxygen in the atmosphere during electrical discharges in storms. The oxides formed are washed to the ground by rain and are converted to nitrates by bacteria.

USES: Great quantities of nitrogen are used in the fixation processes which produce the raw materials for the fertilizer and explosive industries as well as materials to be used in the drug and dye industries. Nitrogen is used as an inert atmosphere in metallurgical operations, in rooms used for the storage of inflammable or explosive materials, and in electric light bulbs to lengthen the life of the filament by preventing its oxidation. Argon is sometimes mixed with nitrogen in filling light bulbs.

PRINCIPAL COMPOUNDS: **Ammonia,** NH$_3$, is a colorless gas with a sharp, irritating odor. It is exceedingly soluble in water: 1176 cc. of it dissolve in 1 ml. of water at 0° C., or 0.9 grams of NH$_3$ dissolve in 1 gram of water. It is made by the Haber process. It is a familiar household cleaning agent, and its vapors are a well-known stimulant. Large amounts of it are used in commercial refrigeration plants in which it chills its surroundings by absorbing energy as it is permitted to expand suddenly. (See Figure 33.) It is the starting material in the manufacture of most other compounds of nitrogen.

Nitric acid, HNO$_3$, is a colorless, volatile liquid with a piercing odor. It is completely miscible with water. It boils at 86° C. and freezes at −47° C. Nitric acid is both an acid and a powerful oxidizing agent. In water solution it is a strong acid. It combines with oxides and hydroxides to form nitrates. Similarly, it forms nitrates

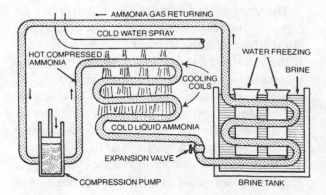

Fig. 33. Use of Ammonia in Refrigeration

with most metals. Nonmetals, such as sulfur or phosphorus, are oxidized by it to sulfates and phosphates respectively. It is made from ammonia by oxidation. The reactions of the **Ostwald process** are:

$$4 NH_3 + 5 O_2 = 6 H_2O + 4 NO \quad \text{(nitric oxide gas)}$$
$$2 NO + O_2 = 2 NO_2 \quad \text{(nitrogen dioxide gas)}$$
$$3 NO_2 + H_2O = 2 HNO_3 + NO$$

Another commercial process for making nitric acid, known as the **arc process,** involves direct combination of oxygen and nitrogen under the influence of an electric arc to form nitric oxide, NO:

$$N_2 + O_2 = 2 NO$$

The nitric oxide is then converted to nitric acid as in the last two steps of the Ostwald process. Nitric acid is used in the manufacture of sulfuric acid, nitrates, fertilizers, dyes, and explosives.

Oxides. Nitric oxide, NO, is a colorless gas. **Nitrogen dioxide,** NO$_2$, is a reddish-brown gas. Both are used in the manufacture of nitric and sulfuric acids. **Nitrous oxide,** N$_2$O, is prepared by heating ammonium nitrate:

$$NH_4NO_3 = 2 H_2O + N_2O.$$

This gas, when inhaled, produces unconsciousness and insensibility to pain. Hence it is used as an anesthetic in minor surgery and dentistry. It is claimed that the inhaling of small amounts of this gas can produce hysterical laughter. Hence it is sometimes called **laughing gas.**

OXYGEN

OCCURRENCE: Oxygen is the most abundant and widely distributed element on earth. 20% of the atmosphere, 50% of the solid crust of the earth, and 89% of water are made up of oxygen. This element is essential to all plant and animal life.

PREPARATION: **Commercially,** oxygen is obtained by the fractional distillation of liquid air. The liquid remaining after the other atmospheric gases have boiled away from liquid air is essentially pure oxygen. It is bottled in tanks under pressure after separation from the other gases. Oxygen is also obtained commercially as a by-product from the industrial electrolysis of water solutions.

In the laboratory, oxygen may be prepared by:

(a) Heating certain metallic oxides, especially the oxides of metals lower than copper in the activity series.

$$2 \, HgO = O_2 + 2 \, Hg.$$

(b) By heating certain oxygen-bearing salts such as potassium chlorate:

$$2 \, KClO_3 = 2 \, KCl + 3 \, O_2.$$

(c) By the reaction between water and sodium peroxide:

$$2 \, H_2O + 2 \, Na_2O_2 = 4 \, NaOH + O_2.$$

PHYSICAL PROPERTIES: These are summarized in Table XXVII.

CHEMICAL PROPERTIES: At room temperature oxygen is only mildly reactive, but at elevated temperatures it combines with most elements and many compounds, especially those containing carbon and hydrogen, to form oxides of all the elements. For example:

(Magnesium)
$2 \, Mg + O_2 = 2 \, MgO.$ (Brilliant white flame)

(Copper)
$2 \, Cu + O_2 = 2 \, CuO.$ (Greenish flame)

(Sulfur)
$S + O_2 = SO_2.$ (Blue flame)

(Alcohol)
$C_2H_5OH + 3 \, O_2 = 2 \, CO_2 + 3 \, H_2O.$ (Yellow flame)

(Carbon tetrachloride)
$CCl_4 + O_2 =$ (No reaction, CCl_4 is nonflammable)

USES: Oxygen is necessary to sustain human life. Cylinders of this gas are carried aboard planes or on mountain-climbing expeditions to permit respiration at high elevations. It makes possible the production of high temperatures in oxyhydrogen and oxyacetylene torches. In hospitals, much oxygen is used to assist in the recovery of patients from lung diseases and pneumonia.

PRINCIPAL COMPOUNDS: The **metallic oxides,** when combined with water, produce hydroxides:

$$CaO + H_2O = Ca(OH)_2.$$

The metallic oxides are therefore known as **basic anhydrides.** The **oxides of nonmetals,** when combined with water, produce acids:

$$SO_2 + H_2O = H_2SO_3.$$

These oxides are therefore known as **acid anhydrides.**

COMBUSTION

A substance combining with oxygen is always oxidized. All such combinations result in the evolution of heat energy. If the rate of reaction is slow, and only heat energy is given off, the process is called **slow oxidation.** But if oxygen combines with the other substance so rapidly that light energy as well as heat is evolved, the process is known as **combustion.** The **flame** produced by combustion consists of burning gases vaporized from the combustible substance by the heat of the reaction. The flame may be colored by energized ions emitting light energy as well as by bits of solid material heated to incandescence by the reaction energy. The rusting of iron is a slow oxidation. The burning of wood is combustion. It should be pointed out that the total amount of energy released by the oxidation of a substance is the same regardless of the rate of the combustion or oxidation process.

Before a substance can burst into flame, it must be heated to a definite temperature. This minimum temperature is known as the **kindling temperature.** Each combustible substance has its own kindling temperature.

EXPERIMENT 25: Select a 9-inch pie tin. On the inside, just at the edge of the sloping side, place at equal space intervals the following items: the head of a match, shavings from a cork, torn bits of paper, a small piece of cotton dipped in oil, a small piece of cotton wet with lighter fluid, a small piece of cotton wet with turpentine. Center the pan directly over the gas burner on the stove and light the gas. The objects in the pan will take fire in the order of increasing kindling temperature.

Spontaneous combustion is likely to occur when the following factors are present:

A **combustible material** which is a **poor conductor of heat** is stored in **still air.**

Oxygen in the air begins slowly to oxidize the combustible material and thereby generate heat. The heat is not conducted away, but is permitted to accumulate around the material. Eventually the temperature is raised to the kindling temperature, and active combustion proceeds from then on. Oily or paint-stained rags should never be permitted to accumulate, especially in the corners of closets or cabinets. If they must be kept, safety dictates that they be stored in metal containers in a well-ventilated spot.

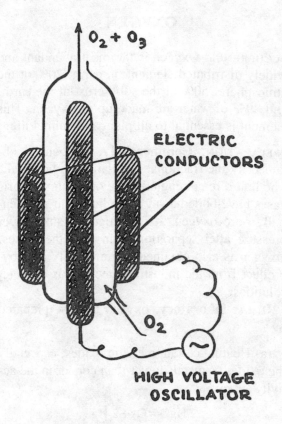

Fig. 34. Ozonizer

OZONE

OCCURRENCE: Ozone, O_3, is an **allotropic form** of oxygen, which means that it is the same substance in different molecular form and possessing different properties. Ozone is found to a slight extent in the atmosphere and in the vicinity of sparking electrical equipment.

PREPARATION: Ozone is prepared by passing oxygen or air through an electrical discharging apparatus known as an ozonizer. (See Figure 34.) About 19% of the oxygen is converted by this process. The equation is:

$$3 O_2 = 2 O_3.$$

PHYSICAL PROPERTIES: These are summarized in Table XXVII.

CHEMICAL PROPERTIES: Ozone, a pale blue gas, is a much more powerful oxidizing agent than ordinary oxygen. Its principal chemical reaction serves as a test for its presence. If air with a trace of ozone is bubbled through a solution of iodide ion, the iodide ion is oxidized to free iodine:

$$O_3 + 2 I^- + H_2O = O_2 + I_2 + 2 OH^-.$$

If starch is also dissolved in the water, an intense blue color will form as the free iodine is liberated.

USES: The uses of ozone depend upon its oxidizing powers. It is a powerful bleaching agent, for it oxidizes colored substances. It is also used to kill bacteria in air and drinking water. Although it has a penetrating odor (you have smelled it around an electric train), it is an excellent deodorizer.

HYDROGEN

OCCURRENCE: Free hydrogen is rare because of the reactivity of this element, but in the combined form, hydrogen makes up about 1% by weight of the crust of the earth. Actually in total number of atoms, hydrogen is probably second to oxygen in abundance on earth, while in the universe it is probably by far the most abundant element. Hydrogen forms about 11% by weight of water, and is to be found in all petroleum products, in all acids and bases, and in all forms of animal and vegetable life.

PREPARATION: **Commercially,** hydrogen is prepared from water. Three different processes are commonly used:

1. The electrolysis of water solutions.
2. The action of steam on hot iron:

$$3 Fe + 4 H_2O = Fe_3O_4 + 4 H_2.$$

3. The **water gas method.** In this method, steam is passed over hot carbon in the form of coke or coal:

$$H_2O + C = CO + H_2.$$

The mixture of hydrogen and carbon monoxide is known as **water gas.** Both of these gases are combustible, so the mixture is used as a gaseous fuel.

In the laboratory, hydrogen is usually prepared by the action of active metals on an acid:

$$Zn + 2 HCl = ZnCl_2 + H_2.$$

It may also be obtained from the action of very active metals on water:

$$Ca + 2 H_2O = Ca(OH)_2 + H_2.$$

PHYSICAL PROPERTIES: These have been summarized in Table XXVII.

CHEMICAL PROPERTIES: Hydrogen is not very reactive at room temperature, but at higher temperatures it burns vigorously, and often explosively, in air or oxygen to form water.

$$2 H_2 + O_2 = 2 H_2O.$$

Hydrogen is a moderately strong reducing agent, as can be seen from its position in Table XXV. For example, it can reduce the oxides of metals less active than manganese to the free metals:

$$H_2 + CuO = H_2O + Cu.$$

Hydrogen will combine directly with the very active metals to form compounds called **ionic hydrides,** in which the valence number of hydrogen is -1.

$$Ca + H_2 = CaH_2.$$

These solids react vigorously with water to give off hydrogen and form the metallic hydroxide.

USES: In the presence of a catalyst and under pressure, hydrogen will combine with vegetable oils to form solid fats like Crisco, which is used as shortening. The process is known as **hydrogenation.** This process is also extensively used in the refining of petroleum products to increase the yields of gasoline. Considerable quantities of hydrogen are used for hydrogenation.

Hydrogen is also used extensively in the production of ammonia and many other chemical substances. It is used as a fuel to propel rocket engines and in fuel cells to generate electricity. Its use in inflating balloons was stopped after the disaster of the zeppelin *Hindenburg*.

PRINCIPAL COMPOUNDS:

1. **Acids** and **bases.** These will be described as the various elements are studied.
2. **Organic compounds.** Chapter 22 will deal with a few of the thousands of such compounds which are known.
3. **Hydrogen peroxide,** H_2O_2. Pure hydrogen peroxide is an oily liquid which freezes at $-0.89°$ C. and boils at $151.4°$. The pure substance is quite unstable, decomposing with violent force when brought in contact with dust or organic material:

$$2 H_2O_2 = 2 H_2O + O_2.$$

It is made by the electrolysis of a concentrated

solution of potassium bisulfate, $KHSO_4$, with a strong current of electricity. This forms the compound potassium peroxydisulfate, $K_2S_2O_8$, which, when heated with steam, forms hydrogen peroxide and regenerates the potassium bisulfate:

$$K_2S_2O_8 + 2 H_2O = 2 KHSO_4 + H_2O_2.$$

The peroxide is then concentrated by distilling it at very low pressure. Dilute solutions of this compound may also be made from the interaction of barium peroxide and ice cold sulfuric acid:

$$BaO_2 + H_2SO_4 = \underline{BaSO_4} + H_2O_2.$$

Hydrogen peroxide of 90% or more concentration is used in liquid-fuel rockets as an oxidizing agent and as a monopropellant. 30% solutions are used as bleaches for various types of fabrics, ivory, feathers, and other substances. 3% solution is used as a disinfectant and mouthwash, and also for mild bleaching of such items as hair.

WATER

OCCURRENCE: By almost any standard, water is the most important chemical compound on earth. It is also the most abundant compound. It is essential to all life processes. Despite the fact that about 75% of the surface of the earth is covered by water, and that water is found in a saturated layer known as the **water table** just below the surface of the land in most areas, lack of water causes vast land areas to be almost uninhabitable, and serious water shortages threaten important agricultural and industrial areas. Two factors contribute to the shortage of water:

1. The extensive solvent action of water renders most of this compound unfit for use. The oceans are vast solutions whose solutes are so highly concentrated that they are poisonous to man and to the plants he raises for food.

2. A staggering percentage of the fresh water man takes for his use is wasted.

About 70% of the human body is water. Animals and plants have correspondingly high percentages of water in their makeup.

PREPARATION: **Commercially,** water is not "prepared" as such. Fresh water from lakes and streams is purified by **sedimentation** and **filtration** to remove the suspended clay, sand, and organic material. Small amounts of chlorine are then added to the water to kill bacteria.

In the laboratory, and for relatively small-scale consumption, water may be separated from its solutes by:

(a) distillation.
(b) deionization.

In the **deionization** process, water is caused to flow through beds of organic **resins.** One type of resin has the property of being able to exchange its own hydrogen ions for all the metallic ions in the water. Another type of resin exchanges its own hydroxide ions for all the nonmetallic ions in the water. Although this process does not remove nonionic solutes, it produces water of relatively high purity. The two types of resins can be regenerated when they have exchanged their capacity of ions, the hydrogen type being regenerated by a solution of sulfuric acid, and the hydroxide type by a solution of sodium hydroxide.

PHYSICAL PROPERTIES: Water is a clear, odorless, liquid which freezes at $0°$ C. and boils at $100°$ C. In many ways it is the most unusual of liquids. That water is a liquid to begin with is a mystery. All other compounds of similar molecular weight and structure are gases at room temperature (H_2S, NO, NO_2, N_2O, NH_3, CH_4, HF, HCl, etc.). Water has the highest specific heat, heat of fusion, and heat of vaporization of any liquid at room temperature. It also has the greatest solvent action. Its temperature of maximum density is $4°$ C., so it expands as the temperature moves in either direction from that figure. Consequently it is the only known liquid that shows expansion on cooling. As a result of the expansion of water at temperatures below $4°$, colder water rises to the surface, and freezing takes place only on the surface of water rather than completely through the liquid. As a result of this, the lower levels of lakes and rivers remain liquid. This protects the various forms of aquatic life.

CHEMICAL PROPERTIES: Water is an essential part of most chemical changes, either as a reactant or as a medium in which reactions take place. Water is an important weak electrolyte, and it reacts with many elements and compounds.

USES: One rather clever use of water might well be pointed out. Many plants, like cranberry plants, cannot survive temperatures below the freezing point of water. To protect these plants, the fields are flooded during the winter when air temperature falls considerably below 32° F. Only the surface of the water freezes, however, and beneath this layer of ice the liquid water, and consequently the plants, are held at temperatures slightly above 32° F.

CARBON DIOXIDE

OCCURRENCE: In addition to forming about 0.03% of the atmosphere, carbon dioxide, CO_2, is dissolved in all natural waters, and is present combined in a variety of carbonate minerals and rocks, the most abundant of which is **limestone,** $CaCO_3$.

PREPARATION: **Commercially,** carbon dioxide is prepared by heating limestone:

$$CaCO_3 = CaO + CO_2.$$

Carbon dioxide is also made by the combustion of coke or natural gas, or by the fermentation of sugars to make alcohol. The gas is collected, compressed to a liquid, and stored in tanks. The other product, CaO, is called **lime,** and it too is commercially important. Carbon dioxide is also a by-product of the fermentation industry. Large amounts of carbon dioxide are added to the atmosphere from the combustion of carbon or carbonaceous material with excess oxygen.

In the laboratory, carbon dioxide is generally prepared by the action of acids on metallic carbonates:

$$2 HCl + MgCO_3 = MgCl_2 + H_2O + CO_2.$$

In nature there is a **carbon cycle,** similar to the nitrogen cycle, tending to keep the amount of carbon dioxide in the atmosphere relatively constant. Plant life, with the aid of sunlight and the catalytic action of **chlorophyll,** can absorb carbon from the carbon dioxide of the atmosphere to form cellulose, sugar, starch, and protein in its cells. Pure oxygen is returned to the air in this process, which is called **photosynthesis.** Animal life eats plants to acquire carbon, and inhales oxygen from the air. Carbon dioxide forms in animal cells, and is exhaled into the atmosphere.

PHYSICAL PROPERTIES: Like water, carbon dioxide has some unusual physical properties. At room temperature it is a colorless, heavy gas. If this gas is cooled to $-79.9°$ C., it will condense directly to a solid. Solid carbon dioxide is known as **dry ice.** Similarly, the solid passes directly to the gaseous state as it is warmed up. This phenomenon is known as **sublimation.** Carbon dioxide must be compressed to at least a pressure of 5.2 atmospheres before any liquid state forms. Above this pressure, the freezing point of carbon dioxide is $-56.6°$ C. Notice that the **normal boiling point is below the freezing point** of this substance! Other physical properties of carbon dioxide are summarized in Table XXVII.

CHEMICAL PROPERTIES: Carbon dioxide is a very stable compound which neither burns nor supports combustion. At high temperatures it can be reduced to carbon monoxide, CO, by hot carbon or zinc:

$$C + CO_2 = 2 CO.$$

It combines with the oxides and hydroxides of the very active metals to form carbonates:

$$CaO + CO_2 = CaCO_3.$$
$$Ca(OH)_2 + CO_2 = CaCO_3 + H_2O.$$

It dissolves in water to form the important weak electrolyte carbonic acid, H_2CO_3.

USES: Large quantities of carbon dioxide are used in the manufacture of white lead, sodium carbonate, and sodium bicarbonate (baking soda). It is also used in the manufacture of carbonated beverages, being dissolved in these liquids under pressure. Removal of the cap from a bottle of soda permits the excess carbon dioxide to bubble free. One type of fire extinguisher contains liquid carbon dioxide under pressure.

Solid carbon dioxide is used as a refrigerant.

NOBLE GASES

OCCURRENCE: **Helium, neon, argon, krypton,** and **xenon** are found in the atmosphere. Helium is also found in natural gas deposits in the southwestern part of the United States. **Radon** is found associated with radium-bearing minerals.

PREPARATION: All except radon are obtained from the fractional distillation of liquid air. Helium is obtained from natural gas deposits by liquefying all other constituents and collecting the gaseous helium. The percentage of helium in these deposits may range from 1% to almost 2%. Radon is radioactive and is a gaseous emanation from radium.

PHYSICAL PROPERTIES: The physical properties of the noble gases are summarized in Table IX on page 23.

USES: Helium is used to inflate balloons and dirigibles. It is added to oxygen to replace the nitrogen in air used by deep sea divers. Nitrogen dissolves in blood under the pressures required for this use, and when the pressure is reduced as the diver emerges, the nitrogen comes out of solution and forms bubbles in the bloodstream. This is believed to be responsible for the painful and sometimes fatal ailment known as "bends." The less soluble helium reduces the possibility of danger from this source. Argon is used with nitrogen in filling electric light bulbs. It, together with helium, is frequently used as an inert atmosphere in scientific work. Neon lights are familiar to all. Table XXVIII supplies some data with regard to the substances used in these lights.

Radon is used in hospitals to combat cancer because of its radioactive properties.

Krypton is used as a gas in incandescent and fluorescent lamps. Xenon is used as a gas in electron and luminescent tubes, flash lamps, and lamps used to excite ruby lasers.

Problem Set No. 14

1. Compute the densities of the substances in Table XXVII relative to Air = 1.
2. The air temperature is 20° C. and the dew point is 10° C. Find the relative humidity.
3. The air temperature on a hot day is 29° C. (84.2° F.). The relative humidity is 89%. Find the dew point (both Celsius and Fahrenheit).
4. On what page of this book is the theory of the use of saturated solutions in air conditioning covered?
5. What is the valence number of sulfur in potassium peroxydisulfate, $K_2S_2O_8$?
6. How many free elements are you likely to inhale in your next breath?
7. 22.4 liters of helium at standard conditions weigh 4 grams. Is the formula for the molecule of helium He, He_2, or He_3?
8. The following statement is made: "When I swing

Table XXVIII
Neon Signs

Color	Gas mixture	Pressure, torr	Color of Glass
White	helium	3–4	clear
Yellow	helium	3–4	amber
Light green	neon-argon-mercury	10–20	green
Dark green	neon-argon-mercury	10–20	amber
Light blue	neon-argon-mercury	10–20	clear
Dark blue	neon-argon-mercury	10–20	purple
Red	neon	10–18	clear
Deep red	neon	10–18	red

my hand rapidly through air, it cools off. If I ride a bicycle fast, I cool off. Therefore, anything that moves through air cools." Comment on the statement, referring to meteorites.

9. Why does a bottle of soda fizz violently after being shaken?

10. Alka-Seltzer tablets contain calcium phosphate, aspirin, citric acid, and sodium bicarbonate. What gas effervesces from the solution when these tablets are dissolved? Why?

SUMMARY

The **atmosphere** is a shell of gases surrounding the earth. It contains nitrogen, oxygen, argon, carbon dioxide, hydrogen, and small amounts of the other noble gases in relatively constant amounts. Water vapor is an important variable component of air.

Relative humidity is the ratio of the partial pressure of the water vapor in the air to the vapor pressure of water at the temperature of the air. The **dew point** is the temperature at which the air would be saturated with water vapor.

Nitrogen is obtained from the atmosphere by the fractional distillation of liquid air. The **fixation** of nitrogen is accomplished in the **cyanamide process,** the **Haber process,** and in the **nitrogen cycle** in nature. Its principal compounds are ammonia, nitric acid, and its oxides.

Oxygen is likewise obtained from the atmosphere by the fractional distillation of liquid air.

It is very active chemically, especially at high temperature. Acid and basic **anhydrides** are among the important compounds of oxygen.

Combustion is oxidation accompanied by the evolution of light energy in the form of a **flame.** The **kindling temperature** of a substance is the lowest temperature at which it will burst into flame. **Spontaneous combustion** may occur if a combustible material which is a poor conductor of heat is stored in still air.

Ozone is an allotropic form of oxygen and is a more powerful oxidizing agent.

Hydrogen is obtained from water or acids. It burns in oxygen and is a good reducing agent. **Hydrogen peroxide** is an important compound of hydrogen used as a source of oxygen for rocket fuels.

Water is the most important chemical compound. Its unique physical properties contribute to its usefulness. It is involved in most chemical reactions either as a reactant or as a medium in which the reaction takes place.

Carbon dioxide is obtained by heating limestone. It participates in the important **carbon cycle** in nature. Solid carbon dioxide **sublimes** because the boiling point of this substance is below its freezing point.

The **noble gases** are usually obtained from air. Helium is found as part of some deposits of natural gas, and radon is a gaseous emanation from radium. Among the noble gases, so far helium, neon, and argon have not been reacted to form compounds.

CHAPTER 13

THE HALOGENS

The elements in Group VII of the periodic table form an important **family** of elements. You will recall that the vertical groups in the periodic table have similarity of electronic structure. Since chemical behavior depends upon electronic structure, we can expect to find similarity of chemical behavior among the elements of a given group. Consequently the groups are frequently referred to as "families of elements."

The elements of Group VII are known as the **halogens,** a term referring to their tendency to form salts. Tables XXIX and XXX, which summarize the physical and chemical characteristics of the elements of this group, should be studied carefully to detect not only similarities, but also the gradual differences that occur as we proceed from the lighter to the heavier elements. The element **astatine,** At, has been omitted from the two tables because it is so rare that it is of no importance. However, by extending the data in the two tables, you should be able to predict fairly well what the properties of this rare element might be.

Note particularly that fluorine is the most active member of the halogen family and that it forms the most stable compounds. It is a general principle of chemistry that **the more active the element, the more stable will be its compounds.** It should also be noted from the replacement properties of these elements that an **activity series of nonmetals** similar to the activity series of metals exists. In order of decreasing activity they are: fluorine, chlorine, bromine, oxygen, iodine, and sulfur. As in the case of the activity series of metals, the more active nonmetallic elements are capable of replacing the less active nonmetals from solutions of the latter's salts. For example,

$$F_2 + 2\,NaBr = 2\,NaF + Br_2.$$

Finally it is very important to keep in mind that all of the halogens and all of the hydrogen halides are **very poisonous** because of their great chemical activity.

Table XXIX
Physical Properties of the Halogens

Characteristic	Fluorine	Chlorine	Bromine	Iodine
Atomic no.	9	17	35	53
Electron arrangement	2, 7	2, 8, 7	2, 8, 18, 7	2, 8, 18, 18, 7
Atomic wt.	19	35.5	80	127
Physical state	gas	gas	liquid	solid
Color	pale yellow	greenish-yellow	dark red	bluish-black
Density, g/cc	1.14 (liq)	1.51 (liq)	3.12 (liq)	4.93 (solid)
Boiling pt., °C.	−188	−35	59	184
Freezing pt., °C.	−220	−101	−7	114
Solubility, g/100 ml water at 0° C.	decomposes to $HF + O_3$	1.46	4.17	−

Table XXX
Table XXX
Chemical Properties of the Halogens

Characteristic	Fluorine	Chlorine	Bromine	Iodine
General activity	Extremely active	Very active	Less active	Least active
Activity with hydrogen	Violent, even in dark	Slow in dark, violent in light	Must be heated	Slow and incomplete even when heated
Formula of hydrogen halide	HF	HCl	HBr	HI
Stability of hydrogen halide	Extremely stable	Very stable	Less stable	Least stable
Oxidizing power	Most powerful	Very powerful	Less powerful	Least powerful
Replacement of nonmetals	Replaces Cl, Br, O, I, S	Replaces Br, O, I, S	Replaces O, I, S	Replaces S only
Reaction with water	Decomposes it to $HF + O_3$	Rapidly forms $HCl + HOCl$	Slowly forms $HBr + HOBr$	No reaction

OCCURRENCE

Because the halogens are so active, none of them is found free in nature. But in the chemically combined state, the halogens are both very abundant and widely distributed. By far the most abundant of these elements is chlorine. Fluorine, bromine, and iodine follow in that order. Table XXXI summarizes the occurrence of the halogens.

PREPARATION

The preparation of free halogens involves vigorous oxidation of compounds containing halogen ions, either by powerful oxidizing or by the oxidizing effect of an electric current. If you refer back to Table XXV you will note that the oxidation-reduction potentials of the halogen half-reactions are:

$2 I^- = I_2 + 2 e^-$	-0.5345 v.
$2 Br^- = Br_2 + 2 e^-$	-1.065 v.
$2 Cl^- = Cl_2 + 2 e^-$	-1.358 v.
$2 F^- = F_2 + 2 e^-$	-2.87 v.

Table XXXI
Occurrence of the Halogens

Element	Occurrence
Fluorine	The mineral fluorite, CaF_2. The mineral cryolite, Na_3AlF_6.
Chlorine	As chloride ion, Cl^-, in seawater (2%). In rock salt beds of the mineral halite, NaCl. In salt beds of KCl, $MgCl_2$, and $CaCl_2$. In human gastric juices as HCl (0.05%).
Bromine	As bromide ion, Br^-, in seawater (0.008%). As NaBr or $MgBr_2 \cdot KBr \cdot 6H_2O$ in salt beds.
Iodine	As iodide ion, I^-, in seawater (0.000004%). As sodium iodate, $NaIO_3$, in Chilean nitrate deposits. In seaweed such as kelp.

Since the fluoride half-reaction is at the very bottom of the list, this means that fluorine is the most powerful oxidizing chemical substance and that no other chemical can oxidize fluoride ions to elemental fluorine gas. Therefore fluorine can be prepared only by electrolysis. Substances such as the dichromate ion, $Cr_2O_7^{--}$, and the permanganate ion, MnO_4^-, as well as fluorine gas can oxidize any of the other three ions because these substances are lower in Table XXV. With the aid of this table you can set up a number of possible reactions which will liberate chlorine, bromine, or iodine from their ionic solutions. Remember that the oxidizing half-reaction proceeds from right to left as written in Table XXV.

Commercially, the halogens are prepared as follows:

1. **Fluorine.** The salt potassium bifluoride, KHF_2, is melted in a copper container fitted with graphite electrodes. Electrolysis of the fused salt liberates fluorine at the anode, and both hydrogen and metallic potassium at the cathode. The latter two elements are valuable by-products. Copper is used as a container because, although it is attacked by free fluorine, a coating of copper fluoride, CuF_2, is produced which forms a protective layer on the metal to prevent further reaction. Lead, nickel, and magnesium have similar behavior toward fluorine and may be used to replace the copper in the electrolysis cell.

2. **Chlorine.** All commercial chlorine now comes from electrolysis. About 90% of commercial chlorine is produced by the electrolysis of **brine,** which is a solution of common salt. (See Figure 35.) In this process, chlorine is collected at the anode and hydrogen gas at the cathode. The resulting solution contains sodium hydroxide, which is obtained as a by-product by the evaporation of the solution. The overall reaction for the process is:

$$2\,NaCl + 2\,H_2O = Cl_2 + H_2 + 2\,NaOH.$$

The electrolysis of fused sodium chloride also yields chlorine gas. The cathode product in this case is a deposit of pure sodium metal. The commercial demand for metallic sodium determines the extent to which this more costly process is used.

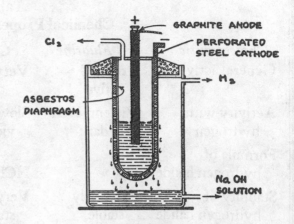

Fig. 35.
Preparation of Chlorine from Salt Solution
in Nelson Cell

3. **Bromine.** Most bromine is now prepared from seawater. The water is first acidified with sulfuric acid, and the bromide ions present are oxidized to elemental bromine by passing chlorine gas into the solution. The reaction is:

$$2\,Br^- + Cl_2 = 2\,Cl^- + Br_2.$$

4. **Iodine.** Most commercial iodine is now obtained from nitrate deposits in Chile, and brines in Michigan and Japan. The iodine ions present are oxidized by a solution of sulfuric acid and sodium nitrite. The reaction is:

$$2\,I^- + 4\,H^+ + 2\,NO_2^- = 2\,H_2O + 2\,NO + I_2.$$

In the laboratory, the halogens may be prepared as follows:

1. **Fluorine.** The laboratory preparation of fluorine is essentially the same as the commercial method.

2. **Other halogens.** The other halogens, chlorine, bromine, and iodine, may be conveniently oxidized from acidified solutions of their ions by the oxidizing action of such ions as the dichromate ion or permanganate ion, or by the action of such compounds as manganese dioxide, MnO_2. Typical reactions are (in these reactions, the symbol X^- represents any of the three halide ions, Cl^-, Br^-, or I^-):

$$6\,X^- + Cr_2O_7^{--} + 14\,H^+ = 2\,Cr^{+++} + 3\,X_2 + 7\,H_2O.$$

$$10\,X^- + 2\,MnO_4^- + 16\,H^+ = 2\,Mn^{++} + 5\,X_2 + 8\,H_2O.$$

$$2\,X^- + MnO_2 + 4\,H^+ = Mn^{++} + X_2 + 2\,H_2O.$$

PHYSICAL PROPERTIES

The major physical properties of the halogens are summarized in Table XXIX. In addition it should be pointed out that all of the halogens form diatomic molecules in which the two atoms are covalently bonded. All have sharp, disagreeable odors and attack the skin and mucous membranes of the nose and throat. Although iodine is a solid at room temperature it readily sublimes because of its high vapor pressure. Iodine vapor is deep violet in color. The solubility of the halogens in non-aqueous solvents such as carbon tetrachloride, CCl_4, and carbon disulfide, CS_2, increase with increased atomic weight. Iodine is about 650 times more soluble in CS_2 than it is in water.

CHEMICAL PROPERTIES

Table XXX summarizes the important chemical properties of the halogens.

EXPERIMENT 26: A very sensitive test for the presence of iodine may be observed as follows. Place a few drops of tincture of iodine in a glass half filled with water. Add a few grains of cornstarch. Stir to dissolve the starch. The blue color that develops is a complex compound of iodine and starch of unknown composition.

USES

1. **Fluorine.** Despite the extremely poisonous nature of both fluorine gas and the fluoride ion, the uses of fluorine are rapidly increasing. Many of its compounds are harmless, such as the important refrigerator gas **Freon,** which is CCl_2F_2. Cryolite, which is sodium aluminum fluoride, Na_3AlF_6, is made synthetically and is a vital flux in the electrolytic production of aluminum metal. Compounds of fluorine and carbon are gaining in importance because of their heat and fire resistance. Many new and important plastics contain fluorine. When drinking water contains about one part per million of fluorine, evidence indicates that such water has a beneficial effect on teeth. However, if the concentration of fluorine in the water increases to more than 3 parts per million, teeth seem to become mottled with brown spots. Fluorine compounds are used as insecticides and wood preservatives. Both lithium and sodium fluorides serve as a flux in the soldering of aluminum.

2. **Chlorine.** Large quantities of chlorine are used to bleach wood pulp for the paper industry and cotton and linen fabrics in the textile industry. Virtually all cities add small amounts of chlorine to drinking water to kill bacteria. Water in swimming pools is likewise chlorinated either directly with chlorine or by adding the compounds calcium hypochlorite, $Ca(OCl)_2$, or sodium hypochlorite, $NaOCl$.

3. **Bromine.** Bromine is used chiefly in the petroleum, drug, and photographic industries. The compound ethylene dibromide, $C_2H_4Br_2$, is an additive in leaded anti-knock gasoline, and a fumigant for grains and fruit. Many dyes and drugs contain bromine. Some bromine compounds have a nerve-soothing effect. Silver bromide is used to coat photographic film and plates.

4. **Iodine.** An alcoholic solution of iodine, known as **tincture of iodine,** is a well-known antiseptic. Other compounds of iodine such as iodoform, CHI_3, are likewise used in the drug industry. Small amounts of iodine are essential in the diet to ensure proper functioning of the thyroid glands. Iodized salt, containing a small amount of sodium iodide, NaI, and seafood are the chief food sources of this element. Deficiency or excess of iodine in the diet leads to the disease called goiter.

PRINCIPAL COMPOUNDS

Hydrogen halides. The hydrogen halides, HF, HCl, HBr, and HI, all exhibit dual behavior. All are covalently bonded gases at room temperature. They are colorless and have penetratingly sharp odors. When perfectly dry, they are nonconductors of electricity. However, they are extremely soluble in water, and in solution they be-

Table XXXII

Physical Properties of Hydrogen Halides

Property	Hydrogen Fluoride	Hydrogen Chloride	Hydrogen Bromide	Hydrogen Iodide
Molecular wt.	20	36.5	81	128
Decomposes at °C.	—	1500	800	180
Density, g/ml, 0°C.	0.00090	0.00100	0.0035	0.00566
Critical temp., °C.	188	51.4	90	150
Critical press., atm.	64	82.1	84.5	81.9
Boiling pt., °C.	19.5	−84.9	−67.0	−35.4
Freezing pt., °C.	−83.1	−114.8	−88.5	−50.8

come ionic substances and dissociate into ions in the manner of all electrolytes. In water solution they are known as the **hydrohalic acids.** Hydrofluoric acid is a weak electrolyte. The others are strong electrolytes. All of the hydrogen halides fume in moist air because they dissolve in the moisture of the air and condense as droplets of acid solution. Table XXXII summarizes the physical properties of the dry hydrogen halides.

The dry halogen halides are relatively inert, but in the presence of even a trace of water they take on their acid characteristics and become vigorously active. Hydrofluoric acid attacks glass and all silicate material, like porcelain. Therefore it must be stored in wax or lead-lined bottles. Hydrochloric acid undergoes all the typical acid reactions with metals and bases. Hydrobromic and hydriodic acids are typical strong acids and are also good reducing agents.

When a fluoride or chloride salt is treated with concentrated sulfuric acid, the more volatile hydrogen halide is formed and passed from the reaction chamber. Typical reactions are:

$$CaF_2 + H_2SO_4 = 2\,HF + CaSO_4$$
$$2\,NaCl + H_2SO_4 = 2\,HCl + Na_2SO_4.$$

However, when bromides or iodides are so treated, they are oxidized by the sulfuric acid completely to the free halogens. The reactions are:

$$2\,NaBr + H_2SO_4 = Br_2 + SO_2 + 2\,NaOH$$
$$2\,NaI + H_2SO_4 = I_2 + SO_2 + 2\,NaOH.$$

Therefore hydrobromic and hydriodic acids must be prepared by an alternative method. The most common method of preparing these acids involves first reacting the halogen with phosphorus to form a phosphorus trihalide, which then reacts with water to produce the acid. For bromine, the reactions are:

$$2\,P + 3\,Br_2 = 2\,PBr_3$$
$$PBr_3 + 3\,H_2O = 3\,HBr + H_3PO_3.$$

The preparation of hydriodic acid would proceed similarly.

The principal use of hydrofluoric acid is in the etching of glass. The area to be etched is first covered with paraffin, which does not react with HF, and the marks or design to be applied is then scratched through the paraffin. The glass is then dipped into the hydrofluoric acid solution, washed, and the paraffin is removed. The glass in electric light bulbs is "frosted" by treatment with this acid.

Hydrochloric acid, known commercially as **muriatic acid,** is one of the most widely used acids. It is essential in the manufacture of textiles, glucose, soap, glue, dyes, and many other chemical substances. It is valuable in removing scale from the surface of metals, and dilute solutions of it are used in a variety of cleaning processes such as the cleaning of mortar from brick or stone walls.

Hydrobromic acid is used in the preparation of

bromides and other chemicals. Hydriodic acid is used both in the preparation of many organic chemicals and as a reducing agent.

Hypochlorous acid. When chlorine gas is dissolved in water, the resulting solution is known as **chlorine water.** A chemical reaction actually takes place as chlorine dissolves. It is:

$$Cl_2 + H_2O = HCl + HOCl.$$

The compound HOCl is called hypochlorous acid. It is a strong oxidizing agent. It is this compound which is responsible for both the bleaching and the disinfecting action of chlorine water, and consequently HOCl must be present during the bleaching process, for dry chlorine does not bleach fabrics. Incidentally, it is the formation of the acids HCl and HOCl which makes chlorine unsuitable as a bleach for silk or wool, for the acids destroy these fabrics.

Hypochlorous acid is a weak electrolyte, so its salts such as sodium hypochlorite, NaOCl, and calcium hypochlorite, $Ca(OCl)_2$, hydrolyze in water solution to form molecules of the acid. Consequently solutions of these salts are effective bleaching agents. Household bleaches such as Clorox and Purex are about 5% solutions of sodium hypochlorite. The oxidizing powers of hypochlorous acid in these solutions cause the germicidal and disinfecting properties of these solutions. Calcium hypochlorite, sold under the trade name of "HTH" (High Test Hypochlorite), is used to disinfect drinking water on ships and water in swimming pools. A related compound, **bleaching powder,** (Ca^{++}, OCl^-, Cl^-), is an effective bleaching agent. It is prepared by passing chlorine gas over moist slaked lime (calcium hydroxide):

$$Ca(OH)_2 + Cl_2 = Ca(OCl)Cl + H_2O.$$

This compound is also known as **chloride of lime.**

Hypochlorous acid solutions and solutions of hypochlorites are stored in brown bottles, preferably in the dark, to avoid decomposition of the acid by sunlight. In direct sunlight, the acid decomposes:

$$2 HOCl = 2 HCl + O_2.$$

Problem Set No. 15

1. Would you expect the halogen astatine, atomic number 85, to be vigorously active chemically or relatively inert?
2. Write a balanced equation for the electrolysis of KHF_2.
3. It is often stated that hydrofluoric acid is much "stronger" than hydrochloric acid because it attacks glass. How would you discuss this matter?
4. What would the anode and cathode products be in the electrolysis of a solution of sodium bromide, NaBr?
5. Some synthetic fibers turn yellow when treated with chlorine bleaches. What would you suspect as the cause of this?

SUMMARY

The **halogens** are the elements in Group VII in the periodic table. They are all active chemically, but decrease in activity as one proceeds down through the group.

The more **active** an element, the more **stable** will be its compounds.

The **activity series of nonmetals** in order of decreasing activity includes: F, Cl, Br, O, I, S.

The halogens are too active to occur free in nature, but their compounds are abundant and widely distributed.

The halogens are **prepared by oxidizing** them, either electrically or by oxidizing agents, from their compounds.

The halogens are **very poisonous.**

The **hydrogen halides** are **covalent** compounds when perfectly dry, and relatively inert. They dissolve even in traces of water to become **ionic hydrohalic acids** which are **vigorously active.** These compounds are likewise **very poisonous.**

The **bleaching** action of chlorine and its compounds is a result of the formation of **hypochlorous acid** in water.

CHAPTER 14

THE SULFUR FAMILY

The elements below oxygen in Group VI of the periodic table are usually called the sulfur family of elements. Of the four—sulfur, selenium, tellurium, and polonium—sulfur is by far the most abundant and most important. All of these elements have six electrons in the outermost shell.

The primary value of studying this group of elements as a family is in gaining an understanding of a change in the type of element encountered as we proceed down through the members of this group. You will recall that the halogens are all nonmetals, but that their activity decreases as we proceed downward through Group VII of the periodic table. In a sense, the halogens become **less nonmetallic** as the atomic weight increases. The steel-gray color of solid iodine gives it a definite metallic appearance. In the sulfur family, where the number of electrons in the outermost shell is fewer than in the case of the halogens, this transition from nonmetallic to metallic characteristics is even more pronounced. Sulfur is a definite nonmetal, although it does show positive valence numbers in many of its compounds just as metals do. Selenium and tellurium, in both appearance and properties, have many metallic characteristics. They are often referred to as **metalloids** because they are so much like metals. Polonium, a very rare radioactive element, is definitely metallic in its properties and behavior.

From a chemical point of view, it is the nature of a nonmetal to gain electrons in its reactions, and substances which behave in that manner are behaving as nonmetals. Substances which lose electrons in their reactions are behaving as metals.

OCCURRENCE

Sulfur. Since sulfur is one of the least active nonmetals at ordinary temperatures, vast amounts of it occur free in nature, particularly in the underground beds of Texas and Louisiana. Impure sulfur is also found in the volcanic regions of Mexico, Japan and Sicily. Chemically combined, sulfur is found as:

(a) Sulfide ores: FeS_2, Cu_2S, PbS, As_2S_3, Sb_2S_3, and Bi_2S_3.
(b) Sulfate ores: $CaSO_4$, $SrSO_4$, and $BaSO_4$.
(c) In organic plant and animal life, particularly substances possessing strong odors and tastes like garlic, onions, horseradish, mustard, and eggs.
(d) In petroleum.

Selenium and **tellurium.** Both of these elements are usually found as impurities in deposits of chemically combined sulfur ores. For example, sulfide ores of copper and lead may be accompanied by small amounts of selenide or telluride compounds such as $PbSe$, Cu_2Se, or $PbTe$. Tellurium is most commonly associated with bismuth as Bi_2Te_3. Selenium is more abundant than tellurium.

Polonium is very rare and may be found in association with other radioactive elements in pitchblende.

PREPARATION

Free sulfur is obtained from the underground deposits by the **Frasch process.** Figure 36 illustrates the concentric pipes inserted into drillings down into the sulfur beds in this process. The steam and hot water in the outer pipe melts the sulfur below, and the hot compressed air forced down through the inside pipe provides the pressure to force the molten sulfur up through the middle pipe. Since the outer and inner pipes are hot, the molten sulfur is kept in the liquid state until it is caused to flow into huge vats where it cools and solidifies. This sulfur is so pure that it needs no further refining for most uses.

Selenium, with some tellurium, forms one of

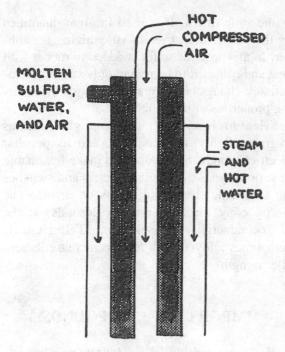

Fig. 36. The Frasch Process

the principal components of the anode slime during the electrolytic refining of copper in a copper sulfate solution. Selenides which form some of the impurities in the copper metal being refined deposit as free selenium on the anode.

Tellurium, in addition, is usually found with bismuth ores. After the bismuth has been removed, the telluride residue is dissolved in HCl solution and then treated with sodium sulfite, Na_2SO_3, solution which oxidizes any telluride present to free tellurium which precipitates from solution.

PHYSICAL PROPERTIES

Table XXXIII summarizes the physical properties of the sulfur family of elements.

In addition it may be pointed out that all of these elements are solid at room temperature, and all are very brittle. Sulfur forms molecules of the formula S_8, while selenium and tellurium form endless chain molecules. Like oxygen, they all have various allotropic forms. Sulfur has three, **rhombic sulfur,** which is stable at room temperature; **monoclinic sulfur,** which is stable between 95.6° C. and the melting point of sulfur; and **amorphous sulfur,** a plastic rubbery form which is obtained by pouring molten sulfur heated almost to the boiling point into water. All forms slowly revert to the rhombic form when allowed to stand at room temperature. Selenium has two principal allotropic forms: **red selenium,** which is amorphous and soluble in carbon disulfide; and **gray selenium,** which is metallic in appearance and insoluble in carbon disulfide. Tellurium also has two principal allotropic forms: **crystalline tellurium,** which is silver white and completely

Table XXXIII
Physical Properties of the Sulfur Family

Property	Sulfur	Selenium	Tellurium
Atomic number	16	34	52
Atomic weight	32	79.0	127.6
Color	Yellow	Red to lead-gray	Black to silver white
Density, g/cc	2.07	4.3-4.8	6.25
Melting point, °C.	112.8	217	452
Boiling point, °C.	444.7	685	1390
Electrical conductivity	Very poor	Poor in dark, Good in light	Good

metallic; and **amorphous tellurium,** which is a brownish-black powder.

CHEMICAL PROPERTIES

When heated in air or oxygen, all of the elements in the sulfur family burn with a blue flame to produce the corresponding dioxide: SO_2, SeO_2, or TeO_2. All other chemical properties are approximately similar among these elements, although the activity decreases slightly as the atomic weight increases. Thus the chemical properties of sulfur serve to illustrate the properties of all the members of this family. These properties may be summarized as follows:

1. Sulfur combines with metals when heated to form sulfides:

$$Zn + S = ZnS \quad \text{(Zinc sulfide)}.$$

Selenides and tellurides are similarly formed.
2. Sulfides react with hydrochloric acid to form hydrogen sulfide:

$$ZnS + 2\,HCl = ZnCl_2 + H_2S.$$

Hydrogen selenide and hydrogen telluride are similarly formed. All three of these compounds are poisonous gases and in solution they are very weak electrolytes. Their solutions are slightly acidic.
3. Sulfur dioxide dissolves in water to form sulfurous acid:

$$SO_2 + H_2O = H_2SO_3.$$

Selenious and tellurous acids are similarly formed. All three of these acids are weak electrolytes. All of them are easily oxidized to form sulfuric acid, H_2SO_4, selenic acid, H_2SeO_4, and telluric acid, H_2TeO_4, respectively.

USES

Sulfur is used as a starting point in many important processes. It is used to make sulfur dioxide, sulfuric acid, matches, black gunpowder, insecticides, and many organic compounds such as the sulfa drugs. It is used in great quantities by the rubber industries in **vulcanizing** the rubber. In this process, sulfur is added to rubber with heat and causes it to be remarkably stable to temperature changes and greatly increases the wearing properties of the rubber.

Selenium is used in making red glass such as in traffic lights, and enamels, and its peculiar electrical conduction properties make it valuable in various types of exposure meters and switches for automatically turning lights on or off. The Xerox copy machine process depends on the photoconductivity of selenium. **Tellurium** is sometimes alloyed with lead to increase its tensile strength.

IMPORTANT COMPOUNDS

Hydrogen sulfide, H_2S. Hydrogen sulfide is a colorless gas at room temperature with the characteristic foul odor of rotten eggs. **This gas is extremely poisonous.** One volume of it in 200 volumes of air can be fatal when breathed for a period of time. Its preparation, by treating sulfides with hydrochloric acid, has previously been mentioned. Its principal use is in analytical chemistry where it is employed to separate mixtures of metallic ions.

Sulfur dioxide, SO_2. Sulfur dioxide is commercially very important. It may be prepared

a. By burning sulfur in air:

$$S + O_2 = SO_2.$$

b. By heating sulfide ores in air. This process is known as **roasting.**

$$2\,ZnS + 3\,O_2 = 2\,ZnO + 2\,SO_2.$$

Roasting not only produces the valuable sulfur dioxide which is collected in the roasting furnaces, but by converting the metallic ore to the oxide, it facilitates the extraction of the free metal from the ore.

c. By the action of strong acids on sulfites. This is the usual method of preparing SO_2 in the laboratory.

$$Na_2SO_3 + 2\,HCl = 2\,NaCl + H_2O + SO_2.$$

Sulfur dioxide is a colorless gas at room temperature with a strong, pungent odor. When dry it is relatively inert. It neither burns nor supports combustion. In the presence of catalysts it will combine with oxygen to form another gas, sulfur trioxide:

$$2 SO_2 + O_2 = 2 SO_3.$$

Both of these oxides of sulfur are acid anhydrides, combining with water as they dissolve in it to form acids:

$$SO_2 + H_2O = H_2SO_3 \quad \text{(sulfurous acid)}$$
$$SO_3 + H_2O = H_2SO_4 \quad \text{(sulfuric acid)}.$$

The most important use of sulfur dioxide is as a starting point in the manufacture of sulfuric acid. But it also has many other uses. When it is liquefied under pressure, it is a valuable solvent used particularly in the refining of lubricating oils. Its water solutions (sulfurous acid) may be used as a bleaching agent for straw, wool, silk, sponges, and other substances which would be injured by chlorine. These solutions are also used to preserve some types of food. Salts of sulfurous acid (sulfites) are used in the manufacture of paper from wood pulp and in the production of the important photographic fixing agent **hypo,** which is sodium thiosulfate:

$$Na_2SO_3 + S = Na_2S_2O_3.$$

EXPERIMENT 27: Sodium thiosulfate is an important reducing agent. This may be observed as follows. Obtain a small package of "hypo" crystals from your photographic supply dealer. A solution will also do. Prepare the blue starch-iodine complex exactly as in Experiment 26. To the blue solution, add a few crystals (or drops) of sodium thiosulfate. Stir. The blue color will disappear because the iodine has been reduced to iodide ion, which does not combine with starch. This reduction has extensive application in analytical chemistry.

Sulfuric acid, H_2SO_4. Sulfuric acid is the most important acid used in chemistry. It is generally prepared by one of two major processes.

a. **The lead chamber process.** Steam, oxygen from air, sulfur dioxide, and oxides of nitrogen which serve as catalysts are introduced into lead-lined chambers where the following reactions take place:

$$2 NO + O_2 = 2 NO_2$$
$$NO_2 + SO_2 + H_2O = H_2SO_4 + NO$$

A coating of lead sulfate forms on the lead linings of the chambers and protects the metal from further attack from the acid. The acid produced by this process is dilute and impure. The lead chamber process now supplies only 3% of the sulfuric acid made.

b. **The contact process.** About 97% of sulfuric acid is made by this process. Filtered and washed sulfur dioxide and air are passed over a catalyst of finely divided platinum or vanadium pentoxide, V_2O_5, at about 400° C. Sulfur trioxide is formed:

$$2 SO_2 + O_2 = 2 SO_3.$$

This gas is then absorbed in concentrated sulfuric acid, forming **fuming sulfuric acid,** or **oleum:**

$$SO_3 + H_2SO_4 = H_2S_2O_7.$$

Water is then added:

$$H_2S_2O_7 + H_2O = 2 H_2SO_4.$$

The product marketed from the contact process has a concentration of about 98%. Concentrated sulfuric acid is a thick, syrupy liquid that boils at about 290° C. and has a specific gravity of 1.84. It is miscible with water in all proportions. When mixed with water, tremendous amounts of heat are liberated. If a small amount of water is dropped into concentrated sulfuric acid, the heat liberated will be sufficient to boil the water and thus cause the acid to be vigorously spattered out of the container. This is most dangerous. Therefore when concentrated sulfuric acid is to be diluted, **SMALL AMOUNTS OF THE ACID ARE ADDED TO THE WATER WITH CONSTANT STIRRING** to dissipate the heat produced.

Chemically, sulfuric acid is both an acid and an oxidizing agent. In dilute solution it is a strong acid, slightly less active than hydrochloric and nitric acids. It reacts with metals above hydrogen in the activity series of metals to liberate hydrogen, and it neutralizes bases. Concentrated sulfuric acid oxidizes both metals and nonmetals, forming sulfates and liberating SO_2. Concentrated sulfuric acid is also an important dehydrating agent or desiccant.

Sulfuric acid, known commercially as **oil of vitriol,** is possibly the most important chemical substance in industry. It is used in the manufacture of hydrochloric and nitric acids:

$$2\ NaCl + H_2SO_4 = Na_2SO_4 + 2\ HCl.$$

$$2\ NaNO_3 + H_2SO_4 = Na_2SO_4 + 2\ HNO_3.$$

It is likewise used in the manufacture of a host of other chemical substances, in the production of phosphate fertilizers, as an oxidizing agent in the refining of petroleum, as an electrolyte in storage batteries, as a bath in electroplating, in the production of explosives, dyes, drugs, synthetic flavors, and paint pigments.

Problem Set No. 16

1. What are the formulas of the following compounds:

    (a) Zinc telluride.　　(c) Selenic acid.

    (b) Hydrogen selenide.　(d) Sodium thiosulfate.

2. Write the balanced equations for the preparation of hydrogen telluride from zinc, tellurium, and hydrochloric acid.
3. What is meant by "allotropic forms" of elements?
4. Which of the following acids are strong and which are weak?

    (a) Sulfurous, H_2SO_3.　　(d) Selenic, H_2SeO_4.

    (b) Sulfuric, H_2SO_4.　　(e) Tellurous, H_2TeO_3.

    (c) Selenious, H_2SeO_3.　(f) Telluric, H_2TeO_4.

5. Which substance is oxidized and which is reduced in the following reactions?

    (a) $S + O_2 = SO_2$.　　(c) $Na_2SO_3 + S = Na_2S_2O_3$.

    (b) $2\ H_2SO_3 + O_2 = 2\ H_2SO_4$.

SUMMARY

The **sulfur family** is composed of the elements below oxygen in Group VI of the periodic table. These elements are less active than the halogens and become more metallic as the atomic weight increases.

Sulfur is by far the most abundant and most important member of this family. The other members in order of decreasing abundance are selenium, tellurium, and polonium.

Sulfur occurs free in nature as well as chemically combined, and is extracted from the underground beds by the **Frasch process.**

All of these elements have several **allotropic forms.**

Hydrogen sulfide is an **extremely poisonous** gas.

Sulfur dioxide is prepared commercially by burning sulfur or by roasting sulfide ores. Its principal use is in the manufacture of sulfuric acid.

Sulfuric acid is the most important manufactured chemical substance. Relatively impure sulfuric acid is made in the **lead chamber process.** Pure sulfuric acid is made in the **contact process.** Safety dictates that **WATER SHOULD NEVER BE ADDED TO CONCENTRATED SULFURIC ACID.**

CHAPTER 15

THE PHOSPHORUS FAMILY

The elements below nitrogen in Group V of the periodic table are known as the phosphorus family of elements. All of the elements in this group have 5 electrons in the outermost shell, and therefore the most common valence numbers exhibited by members of this family are +3 and −5. Phosphorus is the only definite nonmetal in the group. Arsenic is a metalloid. Antimony and bismuth are definite metals.

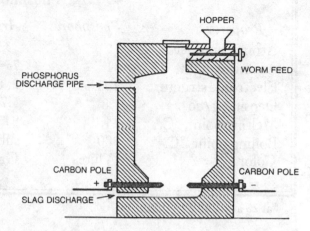

Fig. 37. Preparation of Phosphorus

OCCURRENCE

Phosphorus is too active to be found free in nature. Practically all of it is found combined with oxygen in the phosphate ion, PO_4^{--}. Its principal ores, phosphorite and apatite, contain calcium phosphate, $Ca_3(PO_4)_2$. Phosphorus is also an essential constituent of bones, teeth, muscle, brain, and nerve tissue. The phosphorus in the body needs to be renewed continuously, so such foods as eggs, beans, fish, milk, and whole wheat which contain phosphorus are necessary in the diet. Plants likewise require phosphorus in their tissue. They absorb phosphate from the soil, and when deficiencies of this vital material occur, phosphate fertilizers must be used.

Small amounts of **arsenic** are found free in nature, but normally arsenic is found as the sulfide As_2S_3 in the mineral realgar. **Antimony** likewise occurs as the sulfide in the mineral stibnite, Sb_2S_3, the largest known deposits of which are in the Hunan Province of China. **Bismuth** is the least abundant of the group and is found as the sulfide Bi_2S_3, the telluride Bi_2Te_3, or the oxide Bi_2O_3, and as an impurity in lead, tin, and copper ores.

PREPARATION

Phosphorus is obtained by heating a mixture of calcium phosphate, sand, and coke to a high temperature, usually in an electric furnace. (See Figure 37.) The overall reaction is:

$$Ca_3(PO_4)_2 + 3\ SiO_2 + 5\ C = 3\ CaSiO_3 + 5\ CO + 2\ P.$$

The volatile phosphorus distills off and is condensed under water to protect it from oxidation.

Arsenic and **antimony** are obtained from sulfide ores by first roasting the ore to the oxides and then reducing the oxides by heating with carbon. The reactions for arsenic are:

$$2\ As_2S_3 + 9\ O_2 = 2\ As_2O_3 + 6\ SO_2$$
$$As_2O_3 + 3\ C = 3\ CO + 2\ As.$$

The reactions for antimony are similar. Antimony may also be obtained by the direct reduction of its sulfide by iron:

$$Sb_2S_3 + 3\ Fe = 3\ FeS + 2\ Sb.$$

Bismuth is most commonly obtained as an anode by-product in the electrolytic refining of lead. It may also be reduced from its oxide ore by heating with carbon in a sloping furnace:

$$Bi_2O_3 + 3\ C = 3\ CO + 2\ Bi.$$

The molten bismuth then flows from the furnace into molds.

Table XXXIV
Physical Properties of the
Phosphorus Family

Property	Phosphorus	Arsenic	Antimony	Bismuth
Atomic number	15	33	51	83
Atomic weight	31	75	122	209
Electronic structure	2, 8, 5	2, 8, 18, 5	2, 8, 18, 18, 5	2, 8, 18, 32, 18, 5
Density, g/cc	1.8	5.7	6.7	9.8
Melting point, °C.	44.1	817*	631	271
Boiling point, °C.	280	sublimes	1700	1560
Color	White	Gray	Silver white	Gray white

*at 28 atm and 613 °C.

PHYSICAL PROPERTIES

Table XXXIV summarizes the physical properties of the members of the phosphorus family.

Phosphorus exhibits two principal allotropic forms, white and red phosphorus. **White phosphorus** is a soft, waxy, translucent solid which becomes brittle at 5.5° C. It is crystalline, insoluble in water, but very soluble in carbon disulfide. It is extremely poisonous. It must be stored under water because when it is exposed to atmospheric oxygen for a few moments it catches fire spontaneously. In the presence of light it becomes yellow as a result of the formation of a film of red phosphorus on its surface. Its molecular formula is P_4. **Red phosphorus** is formed by heating white phosphorus in the absence of air at about 250°. It is usually considered amorphous, but it is really a mixture of several other crystalline allotropic forms of phosphorus. It is insoluble in both water and carbon disulfide, and it is not poisonous. It is safe to handle because it does not burn spontaneously at room temperature.

Arsenic has several allotropic forms, the most common being a gray crystalline form with a metallic appearance. It is very poisonous. It cannot be eliminated by the body, so if small quantities are taken internally the arsenic will accumulate until a fatal dose is developed.

Both antimony and bismuth are brittle, crystalline, metallic solids. Both have the property of expanding on solidification just as ice does.

CHEMICAL PROPERTIES

All of these elements combine with oxygen to form oxides. All form the trioxide when burned with a limited amount of oxygen, and all except bismuth form the pentoxide when burned in excess oxygen. For antimony the equations are:

$$4\ Sb + 3\ O_2 = 2\ Sb_2O_3.$$
$$4\ Sb + 5\ O_2 = 2\ Sb_2O_5.$$

The oxides of phosphorus, arsenic, and antimony combine with water to form acids. The acidity of these compounds decreases rapidly with increased atomic weight of the element. Bismuth trioxide combines with water to form a base, bismuth hydroxide. All of these compounds are weak electrolytes. Typical reactions are:

$P_2O_3 + 3\ H_2O = 2\ H_3PO_3$ (Phosphorous acid)

$P_2O_5 + 3\ H_2O = 2\ H_3PO_4$ (Phosphoric acid)

$Bi_2O_3 + 3\ H_2O = 2\ Bi(OH)_3$ (Bismuth hydroxide)

Arsenous, arsenic, antimonous, and antimonic acids are formed in the same manner as phosphorous and phosphoric acids. (Note the difference in spelling between phosphorous acid and the element phosphorus.)

All of these elements combine with hydrogen to form gases similar to ammonia which are extremely poisonous and which possess increasingly repulsive garlic-like odors. These compounds are: arsine (AsH_3), phosphine (PH_3), stibine (SbH_3) and bismuthine (BiH_3).

USES

Most phosphorus is used in making matches. Red phosphorus is used directly, and white phosphorus is used in making tetraphosphorus trisulfide, P_4S_3, which has replaced white phosphorus in the striking tips of matches. White phosphorus is also used in some types of rat poisons, and in the manufacture of military incendiary grenades and bombs. Arsenic is used in many insecticides such as paris green (copper arsenite-acetate, $(CuO)_3As_2O_3 \cdot Cu(C_2H_3O_2)_2$) and acid lead arsenate, $PbHAsO_4$. These are especially effective against leaf-eating insects. It is used in some drugs, improves the quality of brass, and when alloyed with lead, hardens it for use in shot. Antimony is an important ingredient in type and Babbitt (bearing) metal. Alloys containing antimony or bismuth expand on solidifying and thus form sharp castings. Bismuth is used chiefly in making low-melting alloys for use in fuses, sprinkling systems, and fire alarms. Some bismuth compounds are used in drugs.

PRINCIPAL COMPOUNDS

Phosphates. The phosphates of sodium and calcium are by far the most important compounds of this family. Tricalcium phosphate, $Ca_3(PO_4)_2$, present in phosphate minerals and rock, is not suitable as a fertilizer because it is so insoluble. However, this compound is converted to the much more soluble monocalcium phosphate by treatment with sulfuric acid:

$$Ca_3(PO_4)_2 + 2\ H_2SO_4 = Ca(H_2PO_4)_2 + 2\ CaSO_4.$$

The mixture of monocalcium phosphate and calcium sulfate so formed is marketed as "superphosphate" fertilizer.

Sodium forms a series of orthophosphates when sodium hydroxide neutralizes a solution of phosphoric acid.

$$NaOH + H_3PO_4 = H_2O + NaH_2PO_4.$$
(Monosodium phosphate)

$$2\ NaOH + H_3PO_4 = 2\ H_2O + Na_2HPO_4.$$
(Disodium phosphate)

$$3\ NaOH + H_3PO_4 = 3\ H_2O + Na_3PO_4.$$
(Trisodium phosphate)

Monosodium phosphate is a source of hydrogen ions in solution, and this acidic characteristic makes it a valuable ingredient in baking powders. **Disodium phosphate** is used in boilers of ships to soften the boiler water by precipitating calcium and magnesium. **Trisodium phosphate** hydrolyzes extensively to produce an alkaline solution which attacks grease, and thus is used as a cleansing agent. It is also used to soften water because it precipitates calcium from solution.

Sulfides. The use of tetraphosphorus trisulfide in matches has been pointed out. The sulfides of the other members of this family are important pigments. **Arsenic trisulfide**, As_2S_3, is yellow; **antimony trisulfide**, Sb_2S_3, is red; and **bismuth trisulfide**, Bi_2S_3, is black.

Problem Set No. 17

1. Fish has been described as "brain food." Why?
2. Write the balanced equations for the preparation of antimony from stibnite by roasting and reduction with coke.
3. Write balanced equations for the preparation of arsenic acid, H_3AsO_4, from arsenic, oxygen, and water.
4. What are the formulas of the following compounds?

   (a) Arsenous acid.   (c) Antimonic acid.
   (b) Antimonous acid.

5. Baking powders contain sodium bicarbonate, $NaHCO_3$, and monosodium phosphate, NaH_2PO_4. Explain the role of these compounds in the baking process.

CHAPTER 16

CARBON, SILICON, AND BORON

Carbon, silicon, and boron are the remaining nonmetals among the elements. Carbon and silicon are the first two elements in Group IV of the periodic table, and have four electrons in their outermost shell. Boron is the first element in Group III of the periodic table and has three electrons in its outermost shell, thereby making it the only nonmetal possessing less than four electrons in the outermost shell.

OCCURRENCE

Although **silicon** is far less familiar than many of the elements, it is second only to oxygen in abundance. It makes up about 25% of the weight of the crust of the earth, with about 70% of the entire land mass consisting of silicon-bearing rocks. Silicon never occurs free, but always combined with oxygen. There are two principal kinds of siliceous substances. These are:

1. The compound **silicon dioxide,** SiO_2. This compound is found in many forms such as quartz, ordinary sand, sandstone, flint, agate, and the semiprecious stone amethyst.

2. **Silicate rocks.** A tremendous variety of silicate rocks exist in which metallic ions are combined with a vast number of different silicate ions, each containing different proportions of silicon and oxygen. These rocks vary from extremely hard substances like **garnet** to very soft substances like **asbestos.** Other silicate materials are clay, mica, talc, zircon, beryl, feldspar, ultramarine, and zeolite, all of which have commercial importance.

Although **carbon** is much less abundant than many elements, it is one of the most readily available elements. It is found free as the mineral **graphite** and as **diamonds.** Carbon is an essential constituent of all plant and animal life. **Coal** is formed by the gradual decay of plant life, being enriched in carbon as a result of the loss of volatile substances like carbon dioxide and methane, CH_4, during the decay process. The formation of coal proceeds in the following steps:

Dry plant Matter→ 44% C	Dry Peat→ 50% C	Lignite→ 57% C	Bituminous or Soft Coal→ 78% C	Anthracite or Hard Coal 81% C

Petroleum and **natural gas** contain compounds of carbon and hydrogen called **hydrocarbons.** These compounds and their derivatives are so numerous that a special field of chemistry, **organic chemistry,** is devoted to a study of them. Carbon is also found as carbon dioxide in the atmosphere, and as carbonate rock, the most important of which is **limestone,** $CaCO_3$.

Boron is relatively rare. It is never found free, but, like silicon, is always found combined with oxygen in **borates.** The most important source of it is **borax,** sodium tetraborate, $Na_2B_4O_7 \cdot 10\ H_2O$. This compound is found in the desert regions of California. Another compound, **boric acid,** H_3BO_3, is found in the volcanic regions of Italy.

PREPARATION

Both **silicon** and **boron** are relatively difficult to obtain in the free state. Both may be prepared either in the laboratory or commercially by the action of powerful reducing agents on their oxides at high temperature. Equations are:

$$SiO_2 + 2\ C = 2\ CO + Si$$
$$SiO_2 + 2\ Mg = 2\ MgO + Si$$
$$B_2O_3 + 3\ Mg = 3\ MgO + 2\ B.$$

The various common forms of **carbon** are obtained as follows. **Graphite** in massive crystals is mined, or is obtained by heating coke and pitch in furnaces to very high temperatures. Volatiles are driven off and large graphite crystals grow in the furnace. **Charcoal,** which consists of tiny graphite crystals, is prepared by the **destructive distillation** of wood, which means that wood is heated in the absence of air to decompose all the complex carbon compounds in the wood, forming volatile substances and leaving a residue of carbon. The volatiles here include wood alcohol, acetic acid, acetone, turpentine (from pine), and many other valuable by-products. **Bone black,** also consisting of tiny graphite crystals, is prepared by the destructive distillation of bones and other packinghouse wastes. **Coke,** which contains 90–95% graphitic carbon, is made by the destructive distillation of soft coal in huge coke ovens. Coal tar and ammonia are valuable by-products from these ovens. **Carbon black,** or **lampblack,** is made by burning a hydrocarbon fuel like natural gas in a very limited supply of air. The sooty flame strikes a cold surface, depositing graphitic carbon which is then scraped off. **Diamonds** are mined. Gem diamonds are obtained from South Africa, and commercial stones for cutting tools are obtained from South America.

PHYSICAL PROPERTIES

The physical properties of these three elements are summarized in Table XXXV.

Carbon has two allotropic forms, diamond and graphite. **Diamond** consists of transparent, oc-tahedral (eight-sided) crystals which, when pure, are colorless. Impurities may change the color to range from pale blue to jet black. Diamond is the hardest known substance, and is a poor conductor of heat and electricity. Its high refractive index is responsible for the brilliance, or "fire" that it exhibits. **Graphite** crystals are made up of layer after layer of sheets of carbon atoms. It is black in color. The layered structure of graphite makes it very soft, for the sheets slip easily over one another. Graphite is a good conductor of electricity. Both allotropic forms of carbon are insoluble in ordinary solvents. Both are soluble in molten iron to an extent of about 2%. Such solutions, when solidified to room temperature, are known as **steel.**

Silicon occurs only as a crystalline solid, either as massive crystals or as a powder consisting of tiny crystals too small to be distinguished with an optical microscope. It is quite hard and brittle. **Boron** has two allotropic forms, a crystalline variety and an amorphous form. Boron is very hard and brittle.

CHEMICAL PROPERTIES

At ordinary temperatures, **carbon** is quite inert. It will combine with oxygen at moderately high temperatures, either directly with elemental oxygen or with oxygen in oxides. This makes carbon an important reducing agent, particularly for metallic oxides. At extremely high temperatures, those attained normally only in an electric furnace, carbon will react with other elements to form carbides. In all its compounds, carbon forms covalent bonds with other elements.

Chemically, boron is very much like carbon. It combines with oxygen only at moderately high temperatures and with other elements at extremely high temperatures. Its compounds are covalent. Silicon is somewhat more active than the others, being able to combine with the more powerful oxidizing agents and with strong bases at ordinary temperatures. The latter reaction would be:

$$Si + 2\,OH^- + H_2O = 2\,H_2 + SiO_3^{--} \quad \text{(Silicate ion)}.$$

Table XXXV
Physical Properties of Carbon, Silicon, and Boron

Property	Boron	Carbon	Silicon
Atomic number	5	6	14
Atomic weight	11	12	28
Electronic structure	2, 3	2, 4	2, 8, 4
Melting point, °C.	2300	3600	1415
Boiling point, °C.	2550	sublimes	2355
Density, g/cc	2.3	2.25–3.5	2.3

USES

Free carbon has a variety of important uses in each of its forms. Graphite is used as a lubricant in dry sticks, suspended in water, or suspended in oil. Molded mixtures of clay and graphite form the "lead" used in pencils. Harder "leads" contain more clay. Graphite is used as the electrode in dry cells and in electric arc furnaces. It is an important ingredient in stove polish and some paints. Charcoal is extremely porous from the loss of volatiles, and has the property of **absorbing** vast quantities of gas onto its surface. Thus it has important uses in gas masks and in water purification. It is also an important decolorizer in the refining of sugar. It is a common fuel, and is the form of carbon most often used in reducing some metallic ores. It is also used in the manufacture of gunpowder. Bone black is also an important decolorizer. Coke is essential in the reduction of iron from its ore. It is also an important fuel. Carbon black is used in the manufacture of carbon paper, printer's ink, shoe polish, and paint. It is an important additive to rubber used in automobile tires. Diamonds are extensively used as abrasive material, in cutting and drilling tools, and in the preparation of gem diamonds for jewelry.

Silicon has two important uses. It is the backbone of semiconductors known as "chips" that form the logic base for calculators and computers. A silicon chip a quarter inch on a side can hold a million electronic components. Also, silicon is the basis for photoelectric or solar cells which transform light into electricity. Silicon solar cells are used on rooftops of homes, in spacecraft, in railroad signaling equipment, in light buoys, and to supply electricity to remote villages.

Free boron is used in flares to provide a green color; it is also used as an igniter in rockets. Boron filaments are used in aerospace structures. Boron is a dopant in silicon semiconductors to enhance conductivity.

IMPORTANT COMPOUNDS

Oxides. Carbon has two important oxides. **Carbon dioxide** was described in Chapter 12. **Carbon monoxide,** CO, is an important fuel gas. It is formed by heating metallic oxides with carbon:

$$ZnO + C = Zn + CO.$$

The gas is then collected and used as a fuel. It may also be prepared by steam over white-hot coke:

$$H_2O + C = H_2 + CO.$$

The mixture of hydrogen and carbon monoxide is known as **water gas,** and is likewise an im-

Table XXXVI

Fuel Gases

Fuel	Manufacture	CO	H$_2$	CH$_4$	Other Hydro-carbons	CO$_2$	N$_2$	Heating Value, cal/liter
Water gas	Heating steam and coke	40	50	1.2	—	4.4	3.8	2,700
Producer gas	Heating steam, air, and coke	20	21	4	—	6.8	3.8	1,600
Coal gas	Destructive distillation of coal	4.3	44.8	41	6.0	1.1	2.3	6,300
Natural gas	Gas wells	—	—	85	14	—	1.0	10,500

portant fuel mixture. In general, carbon monoxide is formed whenever carbon is burned in a limited supply of oxygen:

$$2\,C + O_2 = 2\,CO.$$

Carbon monoxide is a colorless, odorless, and tasteless gas. It is very poisonous. Breathing one volume of it mixed with 800 volumes of air for one half hour can be fatal. It burns with a blue flame to form carbon dioxide:

$$2\,CO + O_2 = 2\,CO_2.$$

It is an important ingredient in many fuel gas mixtures. Table XXXVI describes some of the important fuel gases.

Silicon dioxide, SiO_2, is frequently referred to as **silica.** It is one of the most abundant natural compounds. When molten silica is cooled, it does not crystallize but instead slowly solidifies to a **glass,** a rigid, undercooled liquid with random arrangement of its atoms. This compound is thus the basic ingredient in the manufacture of various types of glasses. Other oxides may be added to silica to produce variation in the properties of glasses. Table XXXVII gives the composition of some common kinds of glass.

Silica and glass are inert to all acids except hydrofluoric. However, basic substances attack silica, especially at elevated temperatures:

$$SiO_2 + 2\,NaOH = H_2O + Na_2SiO_3$$
$$\text{(sodium silicate).}$$

Sodium silicate is a glass that dissolves in water, and hence it is often called **water glass.** It is used in soaps and cleansers, in fireproofing, as an adhesive, and in preserving eggs. Other silicates form very important minerals, such as asbestos or mica, and silicates containing aluminum are heated with limestone to make **Portland cement.** When soluble silicates are dissolved in dilute acid solutions, a stiff jelly is formed. When this jelly is dehydrated, **silica gel,** a porous variety of silica, is obtained. This substance is an important dehumidifying agent in commercial air-conditioning units and is used by the chemical industry as a catalyst. It likewise is able to adsorb gases as charcoal does, and thus is used to recover industrial vapor.

Boric oxide, B_2O_3, prepared by dehydrating boric acid by heat, is used as one of the important additives in making chemical- and thermal-resistant glasses. **Boric acid,** H_3BO_3, is prepared by heating an acidified solution of borax:

$$Na_2B_4O_7 + 2\,HCl + 5\,H_2O = 2\,NaCl + 4\,H_3BO_3.$$

Table XXXVII
Composition of Glasses

Kind of Glass	SiO_2	B_2O_3	Na_2O	K_2O	MgO	CaO	ZnO	PbO	Al_2O_3	Fe_2O_3
Window	70.6		17		0.1	10.6			0.8	0.1
Bottle	74		17		3.5	5			1.5	0.4
Crown optical	74.6		9	11		5				
Borosilicate optical	68.1	3.5	5	16				7		
Light flint optical	54.3	1.5	3	8				33		
Heavy flint optical	38			5				57		
Pyrex chemical resistance	80.5	12	3.8	0.4					2.2	
Vycor low expansion	96	4								

This compound is a weak electrolyte and is only sparingly soluble in water. It is used as an eye-wash and as an antiseptic. Borates, such as **borax,** hydrolyze to form mildly alkaline solutions. This causes borax to be a good cleansing agent. Because calcium and magnesium borates are insoluble, borax is a good water-softening agent. It is also a mild antiseptic, it is a good flux, and it is used by the glass industry as a source of boron.

Halides. Carbon tetrachloride, CCl_4, is the most important halide of carbon. It is prepared indirectly because carbon and chlorine do not combine directly. Carbon and sulfur combine readily at high temperatures to form **carbon disulfide:**

$$C + 2\ S = CS_2.$$

This liquid is then treated with chlorine gas:

$$CS_2 + 3\ Cl_2 = S_2Cl_2 + CCl_4.$$

Carbon tetrachloride is then distilled from the sulfur monochloride, which is a valuable by-product used in the rubber industry. Carbon tetrachloride is a colorless, noninflammable liquid which is an important industrial solvent, is used in dry-cleaning clothes, and is an important fire extinguisher.

Silicon tetrafluoride, SiF_4, and **silicon tetrachloride,** $SiCl_4$, fume in moist air. They are gases, and when mixed with ammonia are used in making smoke screens and for skywriting.

Boron trifluoride, BF_3, is a gas and is a vital catalyst in the low-temperature manufacture of synthetic rubber.

Carbides. Calcium carbide, CaC_2, is prepared by heating lime with coke in an electric furnace.

$$CaO + 3\ C = CO + CaC_2.$$

It reacts with water at room temperature to produce the important gas **acetylene,** C_2H_2.

$$CaC_2 + 2\ H_2O = Ca(OH)_2 + C_2H_2.$$

This gas is a very important starting material in the manufacture of a vast number of organic chemicals. It is also a valuable fuel in oxyacetylene torches. Buoys marking sea lanes are charged with calcium carbide, which reacts with seawater to produce the acetylene that is periodically ignited to provide flashing lights on these buoys.

Iron carbide, Fe_3C, is present in all steel and is responsible for the hardness and other properties of steel. **Silicon carbide,** SiC, and **tungsten carbide,** WC, are among the hardest known substances. They have important uses as abrasives and in cutting, grinding, and polishing implements.

Hydrogen cyanide. Hydrogen cyanide, HCN, occurs in several plant products, particularly peach kernels and laurel leaves. This compound is a most poisonous gas. It is a weak electrolyte, so cyanide salts hydrolyze to produce this gas. Sodium and potassium cyanides are used in solution as electroplating baths, and molten sodium cyanide is an important quenching medium in heat-treating steels.

Problem Set No. 18

1. What compound would be formed if carbon were burned:

   (a) In a limited supply of oxygen?
   (b) In an excess of oxygen?

2. How would you account for the fact that producer gas has a lower heating value than water gas?
3. How many different elements are present in ordinary window glass?
4. Why is it advisable to caution children not to eat the kernel inside peach stones?
5. How can the elimination of carbon dioxide by volatilization during the coal formation process cause the decaying plant matter to be enriched in carbon?

CHAPTER 17

THE ALKALI METALS

METALS

Of the 108 elements listed, 86 are metals. The rise of civilization is marked by our increased ability to utilize the remarkable properties of metals. Primitive Stone Age cultures advanced through the Bronze Age to the Iron and Steel Age as people discovered methods of winning pure metals from their ores. Only a few metals such as gold, silver, platinum, and copper occur free in nature. These are called **noble metals.** Practically all of the metals are normally found chemically combined in mineral deposits. A mineral deposit containing sufficient combined metal to make it economically possible to extract the metal is called an **ore.** The rock material associated with the ore, usually silicate rock, is called **gangue.**

The extraction of a metal from its ore usually requires several steps. First the ore is **concentrated,** most commonly by some **flotation** process. The crushed ore is placed in a bath containing a **wetting agent,** a liquid which selectively wets the ore but not the gangue, and a **foaming agent,** a solute which forms a stiff foam with the wetting agent on agitation. When the bath containing the ore is agitated, the wetting agent carries the ore up into the foam, leaving the gangue unaffected at the bottom of the bath. By separating the ore-bearing foam from the rest of the bath, the ore is concentrated.

Then the metal is **reduced** from the ore by some suitable process. The extremely active metals must be reduced by **electrolysis,** where they deposit as the pure metal on the cathode of the electrolysis cell. Slightly less active metals may be liberated from their ores by the action of the very powerful **reducing agents** such as metallic aluminum. Metals like zinc, iron, and lead may be reduced from their oxides by **carbon.** If their ores are sulfides or carbonates, such ores are first **roasted** to the oxide prior to reduction with carbon. The least active metals may be **precipitated from solution** by adding a more active metal to the solution.

Table XXXVIII contrasts the general physical properties of metals and nonmetals.

The chemical activity of the metals is summarized in the Activity Series of Metals in Table XXIV on page 89.

Alloys are composed of two or more metals and have metallic characteristics. They are economically more important than pure metals. They

---

Table XXXVIII

Physical Characteristics of Metals and Nonmetals

Metals	*Nonmetals*
1. All except Hg are solids.	1. At least half are gases.
2. Malleable and ductile solids.	2. Brittle solids.
3. Good conductors of heat and electricity.	3. Poor conductors of heat and electricity.
4. Shiny luster; good reflectors.	4. No luster; poor reflectors.
5. Have only a few electrons in outermost shell.	5. Have many electrons in outermost shell.
6. Lose electrons easily to form cations.	6. Gain electrons to form anions.
7. Good reducing agents.	7. Good oxidizing egents.
8. Hydroxides are basic.	8. Hydroxides are acidic.

are usually formed by melting the metals together and permitting the mass to solidify. **Amalgams** are alloys of mercury. **Cermets** are "alloys" of ceramic silicate materials with metals. They combine the strength and toughness of metals with the heat and oxidation resistance of the ceramic material. They are used in high-temperature applications, such as in gas turbines, jet aircraft parts, and in rockets, where the toughness of metals and the temperature resistance of ceramics are required in a single substance. One cermet contains 85% aluminum and 15% aluminum oxide.

THE ALKALI METALS

The alkali metals are the elements of Group I of the periodic table. They are called alkali metals because their hydroxides are strongly alkaline, and because many of their salts hydrolyze to form alkaline solutions. Sodium and potassium are the two most important members of this family. Lithium, rubidium, cesium, and the extremely rare francium are the other members of the group. All have but one electron in their outermost shell. They are extremely active chemically, for they readily give up this electron to form stable ionic compounds.

OCCURRENCE

The alkali metals are much too active to be found free in nature. Sodium and potassium rank sixth and seventh in the list of abundance of the elements. Sodium ion makes up about 1.14% of seawater. Sodium salts are also found in great abundance in desert regions, particularly as $NaNO_3$, Na_2SO_4, Na_2CO_3, and $Na_2B_4O_7$. Vast deposits of rock salt, NaCl, also exist. Many complex silicate rocks also contain sodium. Potassium makes up about 0.04% of seawater. It is found in vast beds of KCl and K_2CO_3. It is even more commonly found than sodium in complex silicate rocks. The other members of the family are quite rare and are normally found only in complex silicate rock or as impurities in sodium and potassium salt beds.

METALLURGY

The alkali metals are almost always prepared by the electrolysis of their fused salts. The metals deposit on the cathode of the cell. In the Downs process, fused salt, NaCl, is the electrolyte used. See Figure 38. The overall equation is:

$$2\,NaCl = 2\,Na + Cl_2.$$

Potassium and the other members of this family are likewise obtained from their molten chlorides.

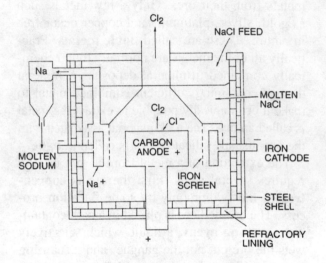

Fig. 38.
The Downs Cell to Manufacture Sodium

PHYSICAL PROPERTIES

The physical properties of the alkali metals are summarized in Table XXXIX.

All are very soft, silvery metals. The softness increases with increased atomic weight. They are all excellent conductors of heat and electricity. When exposed to light, these metals, particularly cesium, have the ability to emit electrons into an evacuated chamber. This makes these metals very useful in photoelectric cells. When they are sufficiently heated they can emit light radiation of characteristic colors. These are:

Li–red   Na–yellow   K–violet
Rb–red   Cs–blue

Table XXXIX
Physical Properties of Alkali Metals

Property	Lithium	Sodium	Potassium	Rubidium	Cesium
Atomic number	3	11	19	37	55
Atomic weight	7	23	39	85.5	133
Electronic structure	2, 1	2, 8, 1	2, 8, 8, 1	2, 8, 18, 8, 1	2, 8, 18, 18, 8, 1
Melting point, °C.	180.5	97.8	63.3	38.9	28.4
Boiling point, °C.	1342	883	760	686	670
Density, g/cc	0.53	0.97	0.86	1.53	1.9

It should be noted that the first three members of this family, lithium, sodium, and potassium, are the only three metals with density less than water.

CHEMICAL PROPERTIES

The alkali metals are intensely reactive. The pure metals tarnish immediately in air, forming a hydroxide film. Their activity increases (they become chemically more metallic) with increased atomic weight. They react so vigorously with water to form hydroxides that large chunks of them dropped into water will cause an explosion. Hydrogen gas is liberated from the water in this reaction. So much heat is generated that potassium burns in contact with water. Cesium will burst into flame in even moist air! These metals must be stored under kerosene, or coated with paraffin, or in sealed, evacuated flasks.

The alkali metals react explosively with dilute acids to liberate hydrogen. They burn vigorously when heated in oxygen or air. In fact, they are so active that they, especially sodium, are often **amalgamated** with mercury to reduce the intensity of their reactions. In the amalgam they retain their chemical properties in a milder degree.

USES

Most of the metallic sodium made is used in the manufacture of tetraethyl lead, an antiknock additive in gasoline. It is also used in the manu-facture of a variety of organic chemicals. Its use in the yellow sodium vapor lights along highways is well known. It is used in the reduction of titanium metal. Alloyed with potassium, it forms an important heat transfer medium in nuclear power plants such as the ones in the Navy's nuclear-powered submarines.

All of these metals are excellent electron sources in photoelectric cells and in cyclotrons.

PRINCIPAL COMPOUNDS

Halides. Sodium chloride, NaCl, is one of the most important chemical compounds. It is mined directly from huge salt beds, or extracted from them by dissolving the salt with water and forcing the brine to the surface with pipes arranged somewhat like those in the Frasch process of extracting sulfur. It is also obtained from seawater by evaporation and recrystallization. This salt is essential in the diet because it is found in all tissue and body fluids. Vast amounts of sodium chloride are used in the preservation of meats and fish and in the preparation of other foodstuffs. Ordinary table salt is highly purified sodium chloride. When perfectly pure it does not absorb moisture, but when slight traces of calcium or magnesium chlorides are present these deliquescent salts cause the table salt to cake up on moist days. Sodium chloride is also used in the manufacture of a variety of other chemicals, in soap, caustic soda, baking soda, and in the glazing of ceramic ware.

Potassium chloride, KCl, is likewise mined from beds in the earth. It is used as an important fertilizer and in the preparation of other potassium compounds. The **bromides** and **iodides** of sodium and potassium are used in medicines and in the photographic industry.

Hydroxides. Sodium hydroxide, also known as **caustic soda,** NaOH, is prepared by the electrolytes of brine (sodium chloride solution). Hydrogen is the cathode by-product and chlorine is the anode by-product. The sodium hydroxide forms in the electrolyte solution in the cell, and is obtained from this solution by evaporation after the electrolysis is complete. The overall reaction is:

$$2\,NaCl + 2\,H_2O = H_2 + Cl_2 + 2\,NaOH.$$

Concentrated solutions of this strong base will injure the skin, and dissolve animal fibers such as wool or silk. Vegetable fibers such as cotton or linen are not attacked by it. Cotton treated with sodium hydroxide takes on a sheen and is known as **mercerized cotton.** Sodium hydroxide is utilized in the manufacture of soap and many sodium compounds. As **lye,** it is a common household cleanser. It is also used in the refining of petroleum.

Potassium hydroxide, KOH, likewise known as **caustic potash,** is prepared electrolytically from a solution of potassium chloride in the same manner as sodium hydroxide. It is used in the preparation of many other potassium compounds and in the manufacture of fine soaps. It is the electrolyte in the Edison storage cell.

Carbonates. Sodium carbonate, Na_2CO_3, and **sodium bicarbonate,** $NaHCO_3$, are the important carbonate compounds of this family. They are prepared by the Solvay process. The raw materials in this process are limestone, ammonia, water, and sodium chloride. The limestone is heated to produce carbon dioxide:

$$CaCO_3 = CaO + CO_2. \qquad (1)$$

The carbon dioxide and ammonia are then bubbled through water, causing the following series of reactions:

$$CO_2 + H_2O = H_2CO_3 \qquad (2)$$

$$NH_3 + H_2O = NH_4OH \qquad (3)$$

$$NH_4OH + H_2CO_3 = H_2O + NH_4HCO_3 \qquad (4)$$

The water is saturated with salt which reacts with the ammonium bicarbonate formed in reaction (4):

$$NH_4HCO_3 + NaCl = NH_4Cl + NaHCO_3. \qquad (5)$$

The sodium bicarbonate precipitates out from the solution, and is filtered from it and dried. The overall reaction for the process is:

$$CO_2 + NH_3 + H_2O + NaCl = NH_4{}^+ + Cl^- + \underline{NaHCO_3}.$$

The dried sodium bicarbonate is then heated strongly to produce sodium carbonate:

$$2\,NaHCO_3 = CO_2 + H_2O + Na_2CO_3. \qquad (6)$$

The Solvay process is remarkable for its lack of waste of materials. The lime formed in equation (1) is added to water to form slaked lime:

$$CaO + H_2O = Ca(OH)_2.$$

When this is added to the filtrate from equation (5) the following reaction occurs:

$$Ca(OH)_2 + 2\,NH_4Cl = CaCl_2 + 2\,NH_4OH.$$

When this is heated, ammonia is recovered from the decomposition:

$$NH_4OH = H_2O + NH_3.$$

CO_2 is recovered from equation (6). Both the CO_2 and NH_3 are reused in the process. Calcium chloride is the only by-product formed.

Sodium bicarbonate, often called **baking soda,** is used in making baking powders and antacid medicinals.

Dry sodium carbonate is known as **soda ash** and is widely used as a source of alkalinity in boiler water and to assist in the softening of water. Soaps, soap powders, cleansing agents, photographic developers and many other products are made from sodium carbonate. Hydrated crystals of sodium carbonate, $Na_2CO_3 \cdot 10\,H_2O$, are known as **washing soda,** and are employed as a common household cleanser.

Nitrates. Both sodium and potassium nitrates are used as meat preservatives, in the manufacture of explosives and fireworks, and as nitrate fertilizers. **Sodium nitrate,** $NaNO_3$, called **Chile saltpeter,** is mined in the desert regions of Chile. It is used to prepare **potassium nitrate,** KNO_3, which is known as **saltpeter.** Saturated solutions of sodium nitrate and potassium chloride are mixed. The least soluble of the four possible salts at low temperatures is KNO_3, which therefore precipitates out. The reaction is:

$$(Na^+, NO_3^-) + (K^+, Cl^-) = (Na^+, Cl^-) + \underline{KNO_3}.$$

Peroxides. When the alkali metals are burned in oxygen or air, the metallic peroxide is formed.

$$2 \, Na + O_2 = Na_2O_2.$$

These peroxides are all powerful oxidizing agents. They are used as bleaching agents and in the preparation of hydrogen peroxide.

OTHER COMPOUNDS

Many other important compounds of sodium and potassium are used industrially. The **chlorates,** like $KClO_3$, are extensively used in the manufacture of fireworks, flares, and matches. **Soap** is fundamentally **sodium stearate,** $NaC_{18}H_{35}O_2$, although potassium stearate is likewise used in fine soaps. **Sodium zeolite,** $NaAlSi_2O_6$, is a very important water softening agent. Other important compounds of sodium and potassium have been described in the chapters dealing with nonmetals.

AMMONIUM COMPOUNDS

The **ammonium ion,** NH_4^+, behaves chemically just like the sodium or potassium ions, combining readily with negative ions in solution to form compounds when the solutions are evaporated. The ammonium ion is formed when ammonia gas dissolves in water to form the weak electrolyte **ammonium hydroxide.**

$$NH_3 + H_2O = NH_4OH.$$

Neutralizing this base with the proper acid will cause the formation of corresponding ammonium salts. For example:

$$NH_4OH + HCl = H_2O + NH_4Cl.$$
(Ammonium chloride)

$$2 \, NH_4OH + H_2SO_4 = 2 \, H_2O + (NH_4)_2SO_4.$$
(Ammonium sulfate)

Many ammonium compounds are industrially important. **Ammonium nitrate,** NH_4NO_3, is rich in nitrogen and therefore is a valuable fertilizer. It is also used in the manufacture of explosives. **Ammonium bicarbonate,** NH_4HCO_3, is a source of CO_2 and is thus used in some types of baking powders. **Ammonium chloride,** NH_4Cl, also called **sal ammoniac,** dissociates into ammonia and hydrogen chloride when hot. This property makes it a valuable soldering flux, for it cleans the surface of the metals being soldered. It is also an important fertilizer, and is used in the manufacture of dry cells.

Problem Set No. 19

1. Write the balanced equation for the preparation of potassium from fused potassium hydroxide.
2. Explain the fact that, unlike the halogens, the alkali metals of higher atomic weight are more active than those of lower atomic weight.
3. What weight of hydrogen gas is evolved when 4.6 g. of sodium metal is added to water?
4. Which raw materials are lost in the Solvay process and what becomes of them?
5. Explain the fact that heating a solution of ammonium hydroxide liberates the gas ammonia.

SUMMARY

Most of the elements are **metals.** Most metals have fewer than four electrons in their outermost shell, so they form compounds through the **loss** of electrons. They are good **reducing agents** and form **basic** hydroxides. They tend to be tough, malleable, and ductile rather than brittle, and they are good conductors of heat and electricity. **Electrometallurgy** involves the reduction of metals from their ores by electrolysis. **Pyrometallurgy** involves the reduction of metals from ores by heating in the presence of a reducing agent. **Hydrometallurgy** involves the precipitation of less active metals from solution by more active ones.

The **alkali metals** are soft, light metals. Having only one electron in their outermost shell, they are **extremely active.** They are always found chemically combined. **Sodium** and **potassium** are among the most common elements, occurring in seawater and in a variety of salt beds. Each of these metals is prepared by the **electrolysis** of a fused compound of the element. The metals have **limited use** because of their great activity, but their compounds are stable and are of great industrial importance.

The **ammonium ion** behaves chemically just like the ions of the alkali metals. Ammonium compounds are likewise important in industry, especially as a source of nitrogen in fertilizers.

CHAPTER 18

THE ALKALINE EARTH METALS AND ALUMINUM

The alkaline earth metals, so named because their oxides form mildly alkaline solutions, are the elements in Group II of the periodic table. The metals in this family all have two electrons in their outermost shell, and readily lose these electrons to form compounds. These elements are only slightly less active than the alkali metals. In fact, the alkaline earths of high atomic weight are more active than the alkali metals of low atomic weight. Calcium and magnesium are the two most important metals in this family.

Aluminum is in Group III just below boron. It has three electrons in its outermost shell which are lost as this metal forms compounds. Its activity is similar to that of the alkaline earth metals, and its physical properties resemble those of magnesium particularly. Therefore it is being considered with the alkaline earth elements.

Beryllium, barium, strontium, and radium are the other metals in the alkaline earth group.

OCCURRENCE

Beryllium is quite rare. Its principal ore is beryl, a complex aluminosilicate, $Be_3Al_2Si_6O_{18}$. When this compound contains traces of chromium impurities, it is the green stone **emerald.** When beryl is contaminated with traces of iron, it is **aquamarine.** Neither beryllium nor any of the other alkaline earth metals occurs free in nature.

Magnesium is the eighth most abundant element, constituting about 2% of the crust of the earth. It makes up about 1.14% of seawater. Its chloride makes up part of the mineral **carnallite,** $MgCl_2 \cdot KCl \cdot 6\ H_2O$, and its sulfate, **Epsom salt,** $MgSO_4 \cdot 7\ H_2O$, is found in beds. It is found as carbonate deposits in **magnesite,** $MgCO_3$, and as **dolomite,** $CaMg(CO_3)_2$. Many complex silicates such as **asbestos,** $CaMg_3(SiO_3)_4$, **talc,** $H_2Mg_3(SiO_3)_4$, and **meerschaum,** $Mg_3Si_2O_5(OH)_4$, also contain magnesium.

Calcium is just behind aluminum and iron in the list of abundant metals, and makes up about 3% of the crust of the earth. It is less abundant in seawater than is magnesium (0.05%), but its land deposits are extensive. Its carbonate, $CaCO_3$, is found in many varieties such as **limestone, marble, chalk, seashells,** and the crystalline mineral **calcite.** It occurs with magnesium in **dolomite. Fluorite,** CaF_2, **anhydrite,** $CaSO_4$, and **gypsum,** $CaSO_4 \cdot 2\ H_2O$, are important calcium minerals. **Apatite,** $Ca_5(PO_4)_3(OH,F,Cl)$, and many complex silicate rocks are also sources of calcium.

The other alkaline earths are relatively rare. **Strontium** is found as the carbonate mineral **strontianite,** $SrCO_3$, and the sulfate mineral **celestite,** $SrSO_4$. **Barium** is found in similar compounds, **witherite,** $BaCO_3$, and **Barite,** $BaSO_4$. **Radium** occurs in minute quantities in uranium ores, the oxide **pitchblende** and the vanadate **carnotite.**

Aluminum is by far the most abundant metal on earth, constituting more than 7% of the crust of the earth. It occurs in a variety of silicate rocks, especially $KH_2Al_3(SiO_4)_3$, **mica,** and $KAlSi_3O_8$, **feldspar.** These rocks disintegrate by the action of moisture and carbon dioxide in the atmosphere to form clay and ultimately aluminum oxide. This process is known as **weathering,** and the principal aluminum ore, **bauxite,** $Al_2O_3 \cdot 2\ H_2O$, has been formed by this process. Aluminum also occurs as **cryolite,** Na_3AlF_6.

METALLURGY

All of the alkaline earth metals may be prepared by the electrolysis of their fused chlorides or by heating the oxides in a vacuum with a powerful reducing agent such as powdered aluminum or silicon (in the form of an alloy with iron known as **ferrosilicon**). Typical reactions are:

$CaCl_2 = Cl_2 + Ca$.   (Electrolysis)

$2\ MgO + Si = SiO_2 + 2\ Mg$.   (Heating in vacuum)

Of the pure metals in this family, magnesium is by far the most abundantly produced, although metallic beryllium, calcium, and barium are commercially important.

Aluminum is prepared electrolytically by the Hall process. (See Figure 39.) Purified alumi-

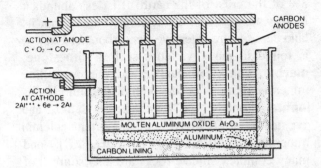

Fig. 39.

Hall Process for the Preparation of Aluminum

num oxide, Al_2O_3, obtained from bauxite, is melted in an electric furnace with the aid of the **fluxes** cryolite and fluorite (fluxes are used to reduce the melting point of a substance). An electric current passed through the molten oxide causes aluminum to be deposited on the carbon cathode lining the furnace. Oxygen is liberated at the carbon anodes immersed in the melt. At the elevated temperature in the furnace, maintained by the heat generated by the resistance of the molten oxide to the flow of current, oxygen attacks the carbon anodes, forming both carbon monoxide and carbon dioxide. Therefore the anodes must be replaced regularly. The fundamental reaction of the Hall process is:

$$2 Al_2O_3 = 3 O_2 + 4 Al.$$

A process that consumes less energy than the Hall process is being developed to make aluminum. Aluminum oxide is treated with chlorine:

$$2 Al_2O_3 + 6 Cl_2 = 4 AlCl_3 + 3 O_2.$$

The aluminum chloride is electrolyzed:

$$2 AlCl_3 = 2 Al + 3 Cl_2.$$

The chlorine liberated by electrolysis is recycled to produce more $AlCl_3$.

PHYSICAL PROPERTIES

The physical properties of the alkaline earth metals and aluminum are summarized in Table XL.

All are silvery white metals and are good conductors of heat and electricity. Note that they are more dense and melt at higher temperatures than the alkali metals. Barium is a good electron emitter like the alkali metals. Calcium, strontium, and barium vapors, when sufficiently hot, emit colored light as follows:

Ca–brick red   Sr–brilliant crimson   Ba–green.

Table XL							
Physical Properties of Alkaline Earths and Aluminum							
Property	*Be*	*Mg*	*Ca*	*Sr*	*Ba*	*Ra*	*Al*
Atomic no.	4	12	20	38	56	88	13
Atomic wt.	9	24	40	88	137	226	27
Electronic structure	2, 2	2, 8, 2	2, 8, 8, 2	2, 8, 18, 8, 2	2, 8, 18, 18, 8, 2	2, 8, 18, 32, 18, 8, 2	2, 8, 3
Melting pt. °C.	1278	649	839	769	725	700	660
Boiling pt. °C.	2970*	1090	1484	1384	1640	1140	2467
Density, g/cc	1.85	1.74	1.55	2.6	3.5	5	2.7

*at 5 torr

Heated ions of the alkali and alkaline earth metals emit the same colors as the heated vapors.

CHEMICAL PROPERTIES

Of the metals in this group, only beryllium, magnesium, and aluminum do not appreciably tarnish on exposure to moist air. All the others form heavy hydroxide coats. A thin film of oxide forms on beryllium, magnesium, and aluminum which then protects these metals from further attack. All of these metals burn with a brilliant white light in air or oxygen. All are good reducing agents, and magnesium and aluminum are extensively used for this purpose. One such reducing reaction involving particularly aluminum is known as the **thermite reaction.** In this reaction, other metals are reduced from their oxides. For example, iron may be reduced by igniting a mixture of powdered aluminum and ferric oxide:

$$Fe_2O_3 + 2\ Al = Al_2O_3 + 2\ Fe.$$

This reaction is highly exothermic, and the heat generated is sufficiently great to melt the iron and permit it to flow into cracks or difficultly accessible spaces. (See Figure 40.) Thus broken iron

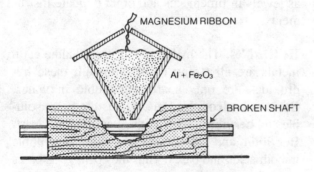

Fig. 40. Thermite Welding

members can be welded in place. The heavier alkaline earth metals combine readily with water at room temperature, liberating hydrogen and forming the metallic hydroxides. Beryllium, magnesium, and aluminum undergo the same reaction at elevated temperatures. All of these metals react vigorously with acid to liberate hydrogen, although aluminum becomes **passive** when treated with concentrated nitric acid because of the formation of an impervious oxide film.

USES

Beryllium is used primarily in alloys, particularly with copper. It greatly hardens the copper. One such alloy, known as a **beryllium bronze,** containing about 2% Be, is used to form "non-sparking" tools. Another is used in the formation of springs of great elastic limit.

Magnesium has become an important structural metal. It is used extensively either in the pure form or alloyed with aluminum in the construction of aircraft and other objects requiring light weight. Magnesium turnings are burned in flashbulbs to produce brilliant light. Powdered magnesium is sometimes used in place of aluminum in the thermite reaction. The other alkaline earth metals have few important uses.

Aluminum has rapidly taken its place as one of the really important structural metals. Its light weight, resistance to corrosion, and tensile strength make it an excellent outside cover for buildings. The fact that it reflects a high percentage of light energy adds to its value as a covering material. Its lightness and strength make it valuable in the manufacture of truck bodies, wheels, railroad equipment and rolling stock, pistons, and a whole host of other objects. Because aluminum is so **malleable** (can be easily rolled into sheets) great quantities of it are utilized in aluminum foil. Because it is so **ductile** (can be easily drawn into wire) much of the wiring in cross-country electrical lines is made of aluminum. Aluminum powder, which consists of tiny flakes of the metal, suspended in a suitable oil medium has extensive use as aluminum paint.

PRINCIPAL COMPOUNDS

Oxides. Oxides of aluminum and the alkaline earth metals all have extremely high melting points. Consequently they are used in the manufacture of firebrick and other refractory materials. **Calcium oxide,** CaO, is known as **lime** or

quicklime, and **magnesium oxide,** MgO, is known as **magnesia.** They are prepared by the thermal decomposition of their carbonates. The process is known as **calcining.** In the case of the preparation of lime, Figure 41, it is also known as the **burning of limestone.** Typical equations are:

$$CaCO_3 = CO_2 + CaO. \qquad \text{(lime)}$$

$$MgCO_3 = CO_2 + MgO. \qquad \text{(magnesia)}$$

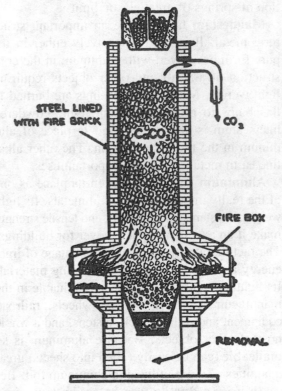

Fig. 41. Preparation of Lime

They both react with water to form sparingly soluble hydroxides. When **barium oxide** is heated, it combines with oxygen to form **barium peroxide.** The equation is:

$$2\,BaO + O_2 = 2\,BaO_2.$$

This compound is sometimes used as a source of hydrogen peroxide:

$$BaO_2 + H_2SO_4 = \underline{BaSO_4} + H_2O_2.$$

In the preparation of lime, the carbon dioxide produced from calcining is a principal source of CO_2 for portable CO_2 fire extinguishers. (See Figure 42.)

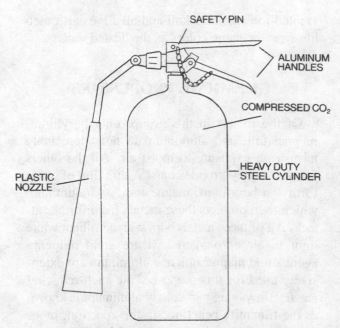

Fig. 42. Carbon Dioxide Fire Extinguisher

Aluminum oxide, Al_2O_3, known as **alumina,** occurs as crystals of **corundum, ruby,** and **sapphire.** It is extremely hard and has widespread use in making abrasive papers and grinding and cutting wheels. Much alumina is utilized in the manufacture of synthetic sapphires and rubies used as jewels in timepieces and other delicate instruments.

Hydroxides. The hydroxides of the alkaline earth metals are all strong bases, although these hydroxides are only sparingly soluble in water. **Barium hydroxide,** $Ba(OH)_2$, is the most soluble and **beryllium hydroxide,** $Be(OH)_2$, the least. Beryllium and aluminum hydroxides are soluble in both acids and bases. Thus they behave as bases toward acids and as acids toward bases. Such hydroxides are said to be **amphoteric.** The amphoteric reactions of aluminum hydroxide are:

As base: $Al(OH)_3 + 3\,HCl = 3\,H_2O + AlCl_3.$

As acid: $H_3AlO_3 + 3\,NaOH = 3\,H_2O + Na_3AlO_3.$

(Sodium aluminate)

Aluminum hydroxide, $Al(OH)_3$, is a sticky, porous material which is quite insoluble in water. It clings to fabrics and adsorbs coloring material which would otherwise not color these fabrics.

Consequently it is an important intermediary in some dyeing processes. When it has been precipitated on cloth it is called a **mordant,** and after it has adsorbed the dye, it is known as a **lake.** Because it adsorbs great quantities of suspended material from solution, it is extensively used in the clarification of municipal drinking water. Since aluminum salts hydrolyze to produce aluminum hydroxide, they may be used as an indirect source of this compound.

Magnesium hydroxide, $Mg(OH)_2$, is known as **milk of magnesia,** a commonly used antacid and laxative.

Calcium hydroxide, $Ca(OH)_2$, is prepared by adding water to lime, and is known as **slaked lime.** It is used in a variety of processes requiring a mild alkali, such as in the removal of hair from hides, in the preparation of mortar and plaster, in water softening, and in the manufacture of many substances such as bleaching powder, paper, glass, and ammonia. **Mortar** is a paste of slaked lime, sand, and water. It sets and dries by absorbing carbon dioxide from the atmosphere, converting the slaked lime to calcium carbonate:

$$Ca(OH)_2 + CO_2 = CaCO_3 + H_2O.$$

The setting of mortar requires much time because the carbonating process proceeds from the outside to the interior of the mortar.

Sulfates. Magnesium sulfate, $MgSO_4 \cdot 7\ H_2O$, when purified, is the commonly used laxative Epsom salt.

Calcium sulfate, $CaSO_4 \cdot 2\ H_2O$, is the important mineral **gypsum.** It is used as an important fertilizer and in the manufacture of a variety of fireproof building materials. It is also used in the manufacture of **plaster of Paris.** When gypsum is heated to about 120° C., it loses 75% of its water of crystallization:

$$2\ CaSO_4 \cdot 2\ H_2O = 3\ H_2O + (CaSO_4)_2 \cdot H_2O.$$
$$\text{(Plaster of Paris)}$$

This reaction is reversed when the proper amount of water is added to plaster of Paris, and the mass quickly sets as crystallized gypsum. A slight expansion accompanies the setting which enables sharp reproduction of details in a mold. The uses of plaster of Paris in wall plaster, for plaster casts,

and in casting and molding operations is well known.

Aluminum sulfate, $Al_2(SO_4)_3 \cdot 18\ H_2O$, is widely used as a source of aluminum hydroxide (by hydrolysis) in the purification of water, in dyeing, and in paper making. It is also used in the preparation of alums. An **alum** is a double sulfate, consisting of the sulfates of a monovalent and a trivalent metal in a single crystalline compound. Their general formula may be written:

$$M^+M^{+++}(SO_4)_2 \cdot 12\ H_2O.$$

A great variety of alums are known, consisting of monovalent sodium, potassium, or ammonium and trivalent aluminum, iron, or chromium. The aluminum alums are used as a source of aluminum hydroxide in essentially the same applications as those of aluminum sulfate.

Calcium carbonate, $CaCO_3$, is an abundant and extremely important mineral. As limestone and marble it is a widely used building material. Marble is the principal stone used by sculptors. Limestone is an important raw material in many manufacturing processes, being a source of both lime and carbon dioxide. Such varied products as mortar, baking soda, and toothpaste require limestone as a raw material. The glass industry uses great quantities of it, and the reduction of iron in a blast furnace requires limestone as a flux. It is likewise an important fertilizer. It is a source of much scenic beauty, from the chalk cliffs of Dover to the magnificent underground limestone caverns found throughout the world. The **stalactites** suspended from the ceilings of these caverns and the **stalagmites** built up from the floors consist of limestone precipitated from saturated solutions.

Cement and concrete. Limestone and clay or shale, $H_4Al_2Si_2O_9$, are heated in a huge rotary kiln forming a complex mixture of lime, alumina, and silica, in the form of calcium aluminosilicates. The clinkered mass is powdered and a small amount of gypsum is added. The mixture is known as **Portland cement.** When this cement is mixed with sand, gravel, and water, it sets to form **concrete.** The setting process involves the formation of crystals of calcium aluminate, calcium hydroxide, and silica. Much heat is evolved, and large batches of concrete must be cooled artificially to ensure proper setting.

Hard water. Water that contains calcium and magnesium ions in solution is known as **hard water** because these ions react with soap, sodium stearate, to prevent the formation of a lather. Thus these ions interfere with the cleansing action of soap.

EXPERIMENT 28: Shave about ⅛ inch of soap from the end of a small bar of white soap. Chop the shavings up very fine and add them to ¼ cup of alcohol. Stir thoroughly to dissolve as much of the soap as possible. Let the excess soap settle. Measure the following into small stoppered jars or bottles: ¼ cup of distilled water, ¼ cup of tap water, ¼ cup of tap water to which 5 crystals of Epsom salt have been added. To each of these liquids in turn, add the clear soap solution drop by drop, shaking the jar well after each addition of soap until a lather is formed which persists for several minutes. The more soap required, the harder the water being tested.

The **softening** of hard water involves the removal of calcium and magnesium ions from solution. If the hard water contains sufficient bicarbonate ion, it may be softened by one of these processes:

(a) By boiling:
$$Ca^{++} + 2 HCO_3^- = \underline{CaCO_3} + H_2O + CO_2.$$

(b) By treating with slaked lime:
$$Ca^{++} + 2 HCO_3^- + Ca(OH)_2 = 2 \underline{CaCO_3} + 2 H_2O.$$

Water containing at least two moles of bicarbonate ion for every mole of calcium or magnesium ion to be removed is called **temporary hard water.** If the water contains less bicarbonate ion, it is known as **permanent hard water.** Such water may be softened by adding soda ash, sodium phosphate, or borax which causes the removal of calcium and magnesium by precipitating them as carbonates, phosphates, or borates respectively.

When large quantities of water are to be softened for industrial purposes, the **ion-exchange** properties of $NaAlSi_2O_6$, sodium zeolite, are frequently utilized. This compound can exchange its sodium ions for calcium and magnesium ions when in contact with hard water, thereby softening it. Then the sodium ions can be returned to the zeolite by treating it with a concentrated solution of sodium chloride. This **regenerates** the zeolite for further use. Synthetic **organic resins** with ion-exchange properties are also widely used for this purpose. An acid resin exchanges hydrogen ions for all metallic ions, and a basic resin exchanges hydroxide ions for all negative ions. The result is the production of **deionized water** of high purity.

Hard water cannot be used in boilers because calcium and magnesium salts will precipitate from boiling water on the walls of boiler tubes forming **boiler scale.** This scale greatly reduces the heat transfer properties of the tubes and may ultimately cause them to blow out. Boiler scale usually consists of calcium sulfate and magnesium carbonates.

Ceramics. Bricks, pottery, chinaware, and porcelain are collectively known as **ceramics.** All of these are made from **clay,** an impure aluminum silicate, or **kaolin,** a purer form of clay. **Bricks** are made by heating a paste of clay and sand until it is thoroughly dehydrated. **Pottery,** consisting only of clay and water, is first shaped on the potter's wheel and then fired in a kiln. It may later be **glazed** by coating it with a silicate glass. **Chinaware** is made from a paste of calcium phosphate, feldspar, and kaolin which is then fired, and later glazed. **Porcelain** is made from quartz, white kaolin, marble, and feldspar. The plastic paste is first molded and then fired.

Problem Set No. 20

1. Carnallite, the double chloride of potassium and magnesium, is melted and then used as a bath for the electrolytic preparation of magnesium.

(a) Will hydrogen, from the water of crystallization, be a by-product? Explain.
(b) Using Table XXV, compute the voltage necessary to decompose KCl and $MgCl_2$ electrolytically.
(c) Suggest a safeguard to ensure the deposition of magnesium free from contamination by potassium.

2. Show with ionization equations how solutions of aluminum sulfate hydrolyze.
3. (a) Write the formula of crystalline sodium aluminum alum.
   (b) Can an alkaline earth metal be present in an alum? Why?
4. Silicate ores are relatively undesirable as a source of a metal, not only because they are rather difficult to decompose chemically but also because of their low metal content. For example, if you found an extensive deposit of pure beryl, what is the maximum amount of pure beryllium you could theoretically extract from the ore?
5. The activity series of metals (page 89) shows only magnesium and calcium from the alkaline earth family. Where would you expect to find strontium, barium, and radium, and in what order?

SUMMARY

The metals of Group II of the periodic table are called the **alkaline earth** metals. Beryllium, magnesium, calcium, strontium, barium, and radium are the members of this family. Aluminum is in Group III, but chemically and physically it resembles the alkaline earth metals.

Aluminum is the most abundant metal, and calcium and magnesium are also very abundant.

The other alkaline earth metals are relatively rare. None of these metals is found free in nature.

These metals are usually prepared by **electrolysis** of a fused compound. The **Hall process** for the electrolytic preparation of aluminum is of great industrial importance. All of these silvery white metals are good conductors of heat and electricity and all are quite active chemically.

Beryllium is used as an alloy with copper to harden the copper. Magnesium and aluminum are important light structural metals.

The oxides of these are used in making refractory materials. Aluminum oxide is an important abrasive. Aluminum hydroxide is **amphoteric** and has adsorptive properties which are widely utilized. Calcium hydroxide is used in many industries both as a source of calcium and as a mild alkali.

The sulfates of these metals tend to be hydrated. Partially dehydrated calcium sulfate is **plaster of Paris.** Aluminum sulfate is the source of many **alums.**

Both calcium and aluminum are involved in the complex silicate structure of **Portland cement.** Calcium and magnesium ions are responsible for the **hardness** of water. **Ceramics** have aluminum silicate as their basic ingredient.

CHAPTER 19

IRON AND THE STEEL ALLOY METALS

It has been said that iron and steel form the skeleton supporting civilization. Pure iron is rarely encountered, but as steel or cast iron it forms the fundamental ingredient of the tools and structures of modern existence. Practically every means of livelihood in some way employs this basic material.

IRON

OCCURRENCE. Iron is second only to aluminum in the list of abundant metals, constituting about 5% of the crust of the earth. The ores containing the greatest concentration of iron are: **hematite,** Fe_2O_3, **magnetite,** Fe_3O_4, and **pyrites,** FeS_2. Hematite is by far the most important of these, occurring in vast but now dwindling deposits in the Mesabi Range in Minnesota. Magnetite is magnetic iron oxide and is known as **lodestone.** Pyrites has a yellow color similar to gold and is often referred to as **fool's gold.** This substance is more commonly employed as a source of sulfur dioxide (by roasting) in the manufacture of sulfuric acid.

Fairly extensive deposits of **siderite,** $FeCO_3$, are also worked for their iron content and a vast amount of low-grade iron ore containing less than 40% iron are abundantly distributed over the earth. This will soon be our chief source of iron. **Meteorites** contain free metallic iron, and it is believed that the liquid core of the earth consists essentially of iron.

METALLURGY. Iron is reduced from its oxide ore in a **blast furnace.** (See Figure 43.) The furnace is **charged** with the following ingredients:

a. Iron ore, consisting of Fe_2O_3 and silicate impurities.
b. Coke.
c. Limestone.
d. Hot air.

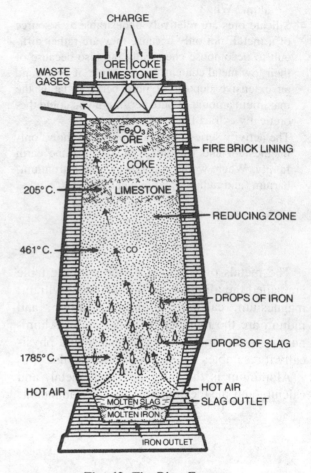

Fig. 43. The Blast Furnace

The solid ingredients are charged at the top of the furnace in layers in the order: limestone, coke, and ore. The hot air is forced upward through the furnace from the bottom. The air oxidizes the coke to carbon monoxide:

$$2\,C + O_2 = 2\,CO.$$

This gas is the reducing agent in the process, and the heat generated in the formation of CO provides the high temperature required for the reduction of the iron:

$$Fe_2O_3 + 3\,CO = 3\,CO_2 + 2\,Fe.$$

The limestone acts as a flux and likewise provides the lime which combines with the silica impurities to form **slag:**

$$CaCO_3 = CO_2 + CaO.$$

$$CaO + SiO_2 = CaSiO_3. \qquad \text{(Slag)}$$

Both the iron and the slag become molten and drop to the bottom of the furnace. The two liquids are immiscible, forming separate layers with the more dense iron at the bottom. Tap holes are provided in the furnace to remove each of these substances from the furnace as liquids.

Solidified iron obtained from the blast furnace is known as **pig iron.** This product is only slightly over 90% pure. It contains two to six percent carbon, appreciable amounts of silicon and manganese, and small amounts of phosphorus and sulfur. The carbon is present as an iron carbide, Fe_3C, known as **cementite,** the compound responsible for the hardness and brittleness of pig iron. If the pig iron is permitted to cool suddenly, all of the carbon is retained as the carbide, and the product is known as **white cast iron.** If the pig iron is cooled slowly, some of the carbon separates from the carbide and forms tiny flakes of graphite. This product is known as **gray cast iron.** Because of the partial decomposition of the carbide, gray cast iron is softer and tougher than white cast iron, it has a higher melting point and it is much more suitable for castings since it **expands** on solidifying, whereas white cast iron **shrinks** on solidifying. Thus gray cast iron is usually used in the making of stoves, radiators, the multitude of other cast objects which are not subject to shock.

If the pig iron is remelted in a furnace with more iron oxide, Fe_2O_3, and stirred or **puddled,** the oxygen from the oxide combines with the carbon to remove it as carbon monoxide. It also combines with phosphorus and sulfur, forming oxides, and with silicon, forming a slag. When the puddling process is complete, the product is rolled or hammered at a temperature just below the solidification point to squeeze out the slag. However, some slag is retained, producing a fibrous structure in the metal. This product is known

as **wrought iron.** It is soft, tough, and relatively corrosion resistant. It is used for making bolts, chains, anchors, fences and grillwork, and iron pipe.

Pure iron is prepared by passing dry hydrogen gas over hot ferric oxide, Fe_2O_3:

$$Fe_2O_3 + 3 H_2 = 3 H_2O + 2 Fe.$$

PHYSICAL PROPERTIES

Pure iron, atomic number 26 and atomic weight 55.8, is a **transition element,** having two incomplete shells of electrons. Its shells contain 2, 8, 14, 2 electrons respectively. Transition elements may use not only the electrons in the outermost shell in forming compounds, but also one or more electrons from the second outermost shell. It will be noted that the transition elements never attain the structure of the noble gases. Iron melts at 1535° C. and boils at 3000° C. Its density is 7.86 grams per cc. It is a relatively soft, silvery white metal, very ductile and malleable, and a good conductor of heat and electricity. It is strongly magnetic.

CHEMICAL PROPERTIES

Iron forms two series of compounds, **ferrous** compounds in which the valence of iron is 2, and **ferric** compounds in which the valence of iron is 3. When pure iron reacts with acids or with steam at high temperature to form hydrogen, the ferrous compounds are formed. These may then be oxidized by atmospheric oxygen or other powerful oxidizing agents to the ferric compounds. The rusting of iron is an electrochemical process which was described in Chapter 11. Dipping iron into concentrated nitric acid, a powerful oxidizing agent as well as an acid, renders iron **passive** as a result of the formation of a film of oxides. Passive iron does not rust unless the film is broken by scratching or striking with a sharp object. Alkalies form a thin film of ferrous hydroxide on iron which then prevents further attack on the metal by the alkalies. Thus iron forms a suitable container for melting sodium and potassium hydroxides in industry.

USES. In addition to the use of iron as a structural material in objects ranging from skyscrapers to toys, its magnetic properties permit its use in magnets, in telephones and telegraph sets, in electric motors and generators, and in many other devices.

COMPOUNDS

Ferric oxide, Fe_2O_3, the principal iron ore, also has many direct uses. It is hard like alumina, and therefore is a valuable abrasive used especially in the polishing of glass. Its red color makes it an important pigment in the paint industry and the cosmetic industry uses it as rouge. **Ferric hydroxide,** $Fe(OH)_3$, is reddish brown in color, but otherwise has essentially the same properties as aluminum hydroxide. Thus it is used as a mordant and in the clarification of drinking water. **Magnetic oxide,** Fe_3O_4, is an equimolar combination of FeO and Fe_2O_3. Besides being an important source of iron, its magnetic properties are utilized in a variety of gadgets. It is formed when iron burns in air or oxygen. Iron **inks** contain a solution of ferrous ions and tannic acid. Oxygen in the air slowly oxidizes the ferrous ions to ferric ions which then precipitate black ferric tannate wherever the ink has been applied. A dye added to the ink shows where writing has taken place until the more permanent ferric tannate precipitates. Since the dye is water-soluble, flushing fresh ink stains with ample water will remove them. But if the stain has been in contact with air for an appreciable time, the ferric tannate must be reduced to remove the stain. A solution of oxalic acid will perform this reduction, regenerating the soluble ferrous ions. Oxalic acid will also "clean" rust stains for the same reason.

STEEL

Steel is fundamentally an alloy of iron and carbon containing less than 2% carbon. If more carbon is present, the alloy is cast iron. The process of making steel involves the removal of carbon from pig iron. Three important processes are in common use today:

a. **The basic oxygen furnace process.**
b. **The open-hearth furnace process.**
c. **The electric furnace process.**

In the **basic oxygen furnace,** Figure 44, the egg-shaped converter is charged with about two-thirds molten pig iron and one-third steel scrap. Fluxes such as lime are added to combine the impurities into a slag. A water-cooled oxygen lance is lowered into the furnace and oxygen is blown onto the top of the metal bath at supersonic speed, creating intense heat of up to 3500° F. within the furnace. The oxygen combines with carbon and other undesired elements, converting the molten mass into steel of quality to be fashioned into automobiles and other products. In less than 45 minutes the cycle is completed, the converter is tilted, and the refined steel is poured into molds.

In the **open-hearth furnace process** (see Figure 45), limestone and equivalent amounts of pig iron and scrap iron are melted in the huge fur-

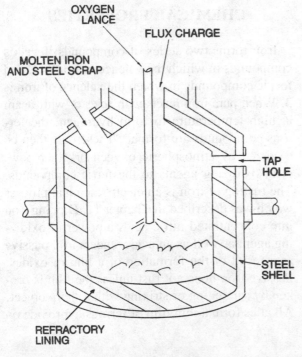

OXYGEN LANCE

FLUX CHARGE

MOLTEN IRON AND STEEL SCRAP

TAP HOLE

STEEL SHELL

REFRACTORY LINING

Fig. 44. The Basic Oxygen Furnace

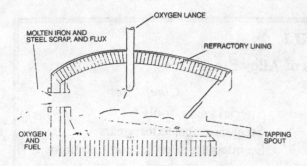

Fig. 45. The Open-Hearth Furnace

nace basin lined with basic magnesia brick. Gaseous fuel is burned with hot air or oxygen over the melt, and the heat is reflected downward onto the molten metal. An oxygen lance also helps combust the gaseous fuel to increase the temperature and hasten the melting. Sulfur and phosphorus are removed by direct combination with the basic lining of the furnace. Since the process requires 8 to 10 hours, there is ample time for careful chemical analysis of the metal being produced. The carbon and manganese contents are controlled by adding sufficient coke and spiegeleisen. When the batch is ready, it is poured into giant ladles and carried away for solidification or for further refinement.

In the **electric furnace process** (see Figure 46), almost perfect control over the composition of the steel can be attained. The charge is mostly scrap iron. The heat is generated from the resistance of molten slag to the flow of electricity.

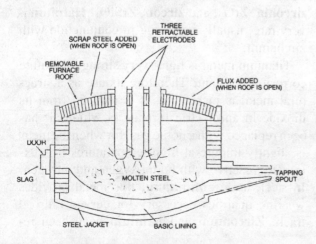

Fig. 46. The Electric Furnace

The current is supplied by three giant carbon electrodes. The basic slag removes the phosphorus and sulfur, and ferrous oxide removes the carbon. The composition is then adjusted by adding the proper ingredients.

Basic oxygen furnace and open-hearth steels are of high quality. They are the fundamental structural steels. Electric furnace steel is the highest quality steel and is used in applications requiring the additional expense of quality such as tools, crankshafts, and gears.

The most effective amount of carbon in hardenable steels may be considered to range between 0.30 and 1.00%. The higher the carbon content, the harder the steel and the more susceptible it is to heat treatment. The following heat-treatment procedures are applicable to most except very low-carbon steel (below 0.06% C):

a. **Annealing.**
b. **Quench hardening.**
c. **Tempering.**

In the **annealing** process, the steel is heated to a bright red color at 1225–1350° F. and then cooled very slowly. This produces a soft form of steel. If, on the other hand, the hot steel is plunged into cold water or oil, the rapid chilling produces a hard form of steel. This is known as **quench hardening.** Hardened steel is brittle because of internal strains. These may be relieved, producing a tougher, although somewhat softer, steel by tempering the hardened steel. This consists of reheating the steel to a temperature below 1200° F. and holding the specimen at the temperature until the stresses are relieved.

Low-carbon steels may be **case-hardened** by pack carburizing—packing the specimen in a mixture of barium carbonate, coke, and charcoal and heating to about 1700° F. for several hours. Carbon is absorbed from the **carburizing mixture** in its outer layers, forming the carbide Fe_3C. This forms a specimen with a very hard shell and a tough interior:

$$BaCO_3 = BaO + CO_2$$

$$CO_2 + C = 2\,CO$$

$$3\,Fe + 2\,CO = Fe_3C + CO_2.$$

Case-hardened steels are also produced by gas

Table XLI
Properties and Uses of Alloy Steels

Alloy Metal	Properties Imparted	Chief Uses
Titanium	Toughness and strength	Special gears; tools
Vanadium	Shock resistance; fine grains	Steel castings; axles; gears
Chromium	Corrosion resistance; hardness and toughness	Stainless steel; tools; cooking utensils
Manganese	Toughness; wear resistance	Safes; steam shovel teeth; crushers; railway frogs and switches
Cobalt	Magnetism	Permanent magnets
Nickel	Hardness; toughness; tensile strength	Automotive parts; armor plate
Molybdenum	Tensile strength; heat resistance	High-speed tools
Tungsten	Hardness; heat resistance	High-speed cutting tools
Silicon	Flexibility; electrical permeability	Springs; transformer cores
Zirconium	Sparking	Lighter flints
Tantalum	Hardness	Tools

carburizing in a furnace with a carbon-bearing atmosphere, or by liquid carburizing in a molten salt bath. Nitriding is a similar process that results in a hard shell containing nitride particles.

ALLOY STEELS

By alloying other metals with steel, a countless number of alloy steels can be produced which possess an almost endless variety of properties and applications. The principal alloying metals used in steel are the transition metals in the same period of the periodic table as iron. They include titanium, vanadium, chromium, manganese, cobalt, and nickel. The metals below chromium in the periodic table, molybdenum and tungsten, are also important steel alloy metals. Table XLI shows the properties and uses of alloy steels.

All of the substances in Table XLI are transition metals except silicon. Locate each of these metals in the periodic table, for we will now examine each one more closely.

TITANIUM, ZIRCONIUM, AND HAFNIUM

Titanium is a very abundant metal, constituting about 0.6% of the crust of the earth. However, it is thinly and fairly evenly distributed over the earth, which makes its exploitation difficult. It occurs as **rutile,** TiO_2, and **ilmenite,** $FeTiO_3$. **Zirconium** is quite rare, occurring as **zirconia,** ZrO_2, and **zircon,** $ZrSiO_4$. **Hafnium** is very rare, usually occurring in conjunction with zirconium.

Titanium metal is light, very strong, and quite corrosion resistant. Therefore its use as a structural metal is increasing. It is reduced from its dioxide in an electric furnace in which air has been replaced with a noble gas, for when the metal is slightly impure at high temperatures it is extremely reactive with both oxygen and nitrogen. In steel, as **ferrotitanium,** this metal provides wearing qualities necessary for curved railroad track. **Zirconium** and **hafnium** have been reduced from their ores only with great difficulty. Both have properties valuable in the exploitation of atomic energy.

Titanium dioxide, TiO_2, is an important white paint pigment. It is used in the making of false teeth, in glazing porcelain, and in whitening paper. A coating of $TiO_2 \cdot Ag \cdot TiO_2$ on the inside of incandescent light bulbs reflects heat back to the filament to generate more light at 30–60% energy savings over standard incandescent light bulbs. **Titanium tetrachloride,** $TiCl_4$, is a fuming liquid used in the making of smoke screens. **Zirconium dioxide,** ZrO_2, is used as a refractory material.

VANADIUM, NIOBIUM, AND TANTALUM

All of these metals are quite rare. **Vanadium** is the most important. It is found in **carnotite,** a uranium vanadate. In the extraction of uranium, vanadium is separated as the pentoxide, V_2O_5. Pure vanadium can then be reduced from the pentoxide by the thermite reaction. **Niobium** and **tantalum** may likewise be reduced from their pentoxides. Tantalum is extremely corrosion resistant. One of the most powerfully corrosive liquids, known as **aqua regia** and consisting of a mixture of one part of concentrated nitric acid and three parts of concentrated hydrochloric acid, will not attack it. As wire, pins, and plates it is useful to surgeons in splicing and supporting bones. **Vanadium pentoxide,** V_2O_5, is an important catalyst in the Haber process for making ammonia and in the contact process for making sulfuric acid.

CHROMIUM, MOLYBDENUM, AND TUNGSTEN

Chromite, $Fe(CrO_2)_2$, is the chief ore of chromium, the most abundant of these three metals. **Molybdenum** is found as the sulfide **molybdenite,** MoS_2, a substance easily mistaken for graphite. Molybdenum is the least common of these

metals. Tungsten occurs in **scheelite,** $CaWO_4$, and in $(FeMn)WO_4$, **wolframite.** Chromium may be prepared by reducing chromic oxide, Cr_2O_3, with aluminum in the thermite reaction, or by depositing it electrolytically from solution. Both molybdenum and tungsten are reduced from their oxides, MoO_3 and WO_3, by hydrogen.

Chromium is a bluish-white, brittle metal that is both hard and corrosion resistant. Most people are familiar with chromium-plated accessories on automobiles and furniture. It is not only one of the most important steel alloy metals, but it is also found in other important alloys such as **nichrome,** an alloy containing chromium, nickel, iron, and manganese, which is used in heating elements in toasters, irons and ovens; and **stellite,** an alloy of cobalt, chromium, tungsten, and carbon, used in surgical instruments, cutlery, and high-speed tools. The chief use of molybdenum is in steel. Tungsten has the highest melting point of all metals (3410° C.). The exceedingly rare metal **rhenium** melts at 3180° C. Tungsten wire drawn from hot compacted powdered tungsten is used as filaments in light bulbs. Both tungsten and molybdenum are used as filaments in electron tubes and electric light bulbs.

Chromium compounds are brilliantly colored, and many of them are used as paint pigments. Table XLII lists some chromium pigments.

Sodium dichromate, $Na_2Cr_2O_7$, is used in the tanning of leather. **Potassium dichromate,** $K_2Cr_2O_7$, is a powerful oxidizing agent in acid solution. **Ammonium molybdate,** $(NH_4)_6Mo_7O_{24} \cdot 4\,H_2O$, is an important reagent in water analysis. **Sodium tungstate,** $Na_2WO_4 \cdot 2\,H_2O$ is used in fireproofing fabrics and cellulose.

MANGANESE

Manganese is quite abundant, although the United States has only limited deposits. However, millions of tons of **manganese nodules** lie on the ocean floor. Their composition is about 28% manganese, 1.4% nickel, 1.2% copper, and 0.25% cobalt. Ocean seabed mining will someday recover these nodules and minerals.

Table XLII
Chromium Pigments

Compound	*Formula*	*Color*
Chromic oxide	Cr_2O_3	Green
Zinc chromate (basic)	$ZnCrO_4 \cdot Zn(OH)_2$	Yellow
Lead chromate	$PbCrO_4$	Yellow
Basic lead chromate	$PbCrO_4 \cdot PbO$	Red
Mixed lead chromate	$PbCrO_4 + PbCrO_4 \cdot PbO$	Orange

On dry land, manganese occurs mainly as the dioxide ore **pyrolusite,** MnO_2. The two metals below it in the periodic table are exceedingly rare. **Technetium,** Tc, and **rhenium,** Re, are their names. Technetium is not found at all in nature, although it has been produced from nuclear reactions. Rhenium is found as a trace impurity in molybdenum and tungsten ores.

Manganese is prepared by reducing the dioxide with carbon in an electric furnace or with aluminum in the thermite reaction. It is a silvery metal with a reddish tint, and is hard and brittle. It is considerably more active than the rest of the transition metals, easily decomposing steam to liberate hydrogen. It is an essential constituent in all steel not only because of the toughness it imparts, but also because it acts as a deoxidizer in the manufacture of steel.

Manganese dioxide, MnO_2, is a powerful oxidizing agent. It is used to decolorize glass and as a depolarizer in dry cells. **Potassium permanganate,** $KMnO_4$, is a deep purple compound which is a powerful oxidizing agent in acid solution. It is used as a disinfectant. A mixture of **manganous sulfate,** $MnSO_4$, and borax is used as a drier in paints.

COBALT AND NICKEL

Both cobalt and nickel are relatively abundant, especially in Canada. **Cobalt** occurs chiefly as the arsenide ore **smaltite,** $CoAs_2$, which is associated with the **nickel** ore **pentlandite,** $(Ni,Fe,Cu)S$. Both cobalt and nickel ores are roasted to the oxide and then reduced in a blast furnace. Cobalt is refined from the blast furnace product electrolytically. Nickel is refined by the remarkable **Mond process.** In this process the gas carbon monoxide is passed under pressure over impure nickel at 100° C. forming the gaseous compound **nickel carbonyl:**

$$Ni + 4 CO = Ni(CO)_4.$$

Thus the nickel is literally wafted away from its impurities. The gaseous carbonyl is then led off to a new chamber where the temperature is raised to 200° C., which causes it to decompose and precipitate pure metallic nickel:

$$Ni(CO)_4 = 4 CO + Ni.$$

The carbon monoxide is then used again to refine more impure nickel.

Cobalt is a silvery metal with a bluish tint. Like iron, it is ferromagnetic. Alloyed in steel, it is used in making permanent magnets. **Carboloy,** an extremely hard alloy used in high-speed cutting operations, consists of tungsten carbide bonded with metallic cobalt. The alloy **alnico,** containing aluminum, nickel, and cobalt (note symbols in name), is one of the most powerfully magnetic substances.

Nickel is likewise ferromagnetic, and is a hard white metal which resists tarnish. It is one of the most important steel alloy metals. As nickel plate it protects steel, copper, and brass from corrosion. When finely divided it is an excellent catalyst, especially in the hydrogenation of fats and oils. The nickel coin contains 25% nickel and 75% copper. Many other nickel alloys are important, especially **monel** metal used in kitchen and bathroom fixtures. It is made of nickel, copper, and iron.

Cobalt salts, such as **cobaltous chloride,** $CoCl_2$, readily form hydrates. When anhydrous, the cobalt ion is blue, and when hydrated it is pink. Thus porous paper or cloth impregnated with cobalt chloride becomes blue in dry weather and pink in rainy weather. Nickel salts are brilliant light green in color. **Nickel dioxide,** NiO_2, is the cathode in the Edison storage battery.

Problem Set No. 21

1. Which ore of iron contains the greatest percentage of iron?
2. How does iron rust?
3. Match the following steel objects with the type of heat treatment most likely used in their preparation:

   (a) File   (1) Tempering
   (b) Razor blade   (2) Case hardening
   (c) Gears (teeth)   (3) Quench hardening.

4. How may ink stains be removed?
5. Write balanced equations for the thermite reduction of:

   (a) Chromium.
   (b) Vanadium.

SUMMARY

Iron is the most important metallic element and is second to aluminum in abundance. Hematite, Fe_2O_3, its principal ore, is reduced to **pig iron** in a blast furnace. Solidified pig iron is **cast iron.**

Steel is made from pig iron by reducing the carbon content to less than 2%. It may be made by the **basic oxygen furnace process,** the **open hearth furnace process,** or by the **electric furnace process.** Over 60% of steel is made by the basic oxygen furnace process. It is the most rapid method. The finished steel may then be **heat-treated** to improve its properties.

Chemically, iron is typical of the **transition elements** which show variable valence because they have two incomplete shells of electrons. Iron forms two series of compounds showing valence numbers of 2 and 3. It alloys primarily with other transition elements in **alloy steels,** substances which show a tremendous variety of properties and uses.

Titanium is becoming an important structural metal, and its dioxide is an important white pigment. **Chromium** is present in stainless steel and is extensively used as a plating on other metals. Many of its compounds are important pigments because of their brilliant color. **Tungsten** is a constituent in the finest high-speed tool steels and is used as filaments in light bulbs. **Manganese** is essential in all steel as a deoxidizer. **Cobalt** improves the magnetic qualities of steel. **Nickel,** refined by the **Mond process,** is present in corrosion-resistant alloys and heating elements in toasters.

CHAPTER 20

NONFERROUS ALLOY METALS

Copper, tin, zinc, and lead have been important metals since civilization began. Copper is second to iron in importance among the metals. These four metals are the principal constituents of a great variety of alloys indispensable in modern civilization.

COPPER

Copper is a fairly abundant metal. It occurs free to some extent, but most copper is extracted from its ores **chalcocite,** Cu_2S; **chalcopyrite,** $CuFeS_2$; and **cuprite,** Cu_2O. It also occurs as basic carbonates in green **malachite,** $Cu_2(OH)_2CO_3$, and $Cu_3(OH)_2(CO_3)_2$, blue **azurite.**

The sulfide ores are first concentrated by **flotation** and then roasted to the oxide before the reducing process begins. The roasted ore is then reduced with coke and air, forming **blister copper** which is about 99% pure. This product is then refined electrolytically (see Figure 47) by using the impure copper as the anode, and thin sheets of pure copper as the cathode in a solution of copper sulfate. The copper in the impure anode dissolves during electrolysis and deposits on the pure cathode. The metals less active than copper in the anode fall to the bottom of the cell as a sludge and are recovered as by-products. The metals more active than copper pass into solution with the copper, but do not deposit on the cathode because copper ions are more easily reduced. Thus, the electrolytic refining of copper, using copper electrodes, takes full advantage of the electrochemical properties of metals and their ions.

Copper is a soft, extremely ductile and malleable metal with a characteristic reddish-brown color. It is easily formed into wire, tubes, and sheets. It is an excellent conductor of electricity and is the metal most often used for that purpose. Since copper is below hydrogen in the activity series of metals, it is not very active chemically,

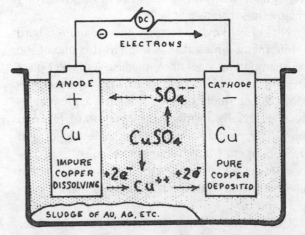

Fig. 47. Electrolytic Refining of Copper

but oxidizing acids react vigorously with it. When exposed for some time to moist air it becomes coated with the green basic carbonate **malachite.** This then protects the metal from further corrosion. It is extensively used for electric wires and cables, drinking water pipes, and in cooking utensils. Its beauty makes it much used for ornamental purposes.

Copper alloys were the metals most used before the advent of steel. Alloys of copper and zinc are known as **brass.** Cartridges, musical instruments, hardware, and various types of pipes

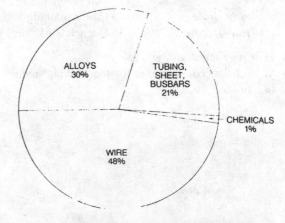

Fig. 48. Uses of Copper

are made from brasses. **Bronze** contains copper, tin, and zinc. It is used for statues and medals, and primitive civilizations used it for tools. **Aluminum bronze** is used for forgings, bolts, and gears. **Beryllium bronze** is used for non-sparking tools. **Bell metal** is made of copper and tin. **Sterling silver** contains $7\frac{1}{2}\%$ copper.

The **nickel** coin is a homogeneous alloy containing 75% copper and 25% nickel. The **dime** and the **quarter** are **clad-metal coins.** The outside layer is an alloy of 75% copper and 25% nickel. The core is copper. If the coins were melted, the composition would be 91.7% copper and 8.3% nickel.

The **half dollar** is also a clad-metal coin. The outside layer contains 80% silver and 20% copper. The core is copper. If the coin were melted, the composition would be 60% copper and 40% silver.

The **silver dollar** minted today is a clad-metal coin as well. The outside layer is an alloy of 75% copper and 25% nickel. The core is copper. Averaging the total metal content, the **Eisenhower dollar** contains 91.7% copper and 8.3% nickel. The **Susan B. Anthony dollar** contains 87.5% copper and 12.5% nickel.

White gold and **dental gold** both contain copper. These are but a few of the many important copper alloys.

Copper is a transition metal and shows valence numbers of 1 and 2 in its compounds. The latter valence number is the more common. **Copper sulfate,** $CuSO_4 \cdot 5\ H_2O$, is the most important copper compound. Known as **blue vitriol,** it is used as a fungicide in water reservoirs and swimming pools. It is used as a bath for copper plating and in the refining of metallic copper.

ZINC, CADMIUM, AND MERCURY

An adequate supply of **zinc** is extracted from its ores: **sphalerite,** ZnS; **smithsonite,** $ZnCO_3$; and **zincite,** ZnO. **Cadmium** is quite rare, but is always present as an impurity in zinc ores. Some **mercury** is found free in nature, but it is normally found as the sulfide ore **cinnabar,** HgS.

Zinc ores are concentrated, roasted to the ox-

ide, and then reduced by heating with carbon. The impure metal obtained by this process contains cadmium. It is refined by distillation with the more volatile cadmium coming off first. Mercury is so volatile that it distills off as its sulfide ore is being roasted. Its vapors are collected and condensed.

$$HgS + O_2 = SO_2 + Hg.$$

Zinc and cadmium are both silvery white metals. Zinc melts at 420° C. and boils at 907° C. Cadmium melts at 320° C. and boils at 767° C. Both are above hydrogen in the activity series of metals, with zinc being more active. Surface oxide films protect them from further corrosion. Zinc hydroxide is amphoteric, but cadmium hydroxide is always basic. Mercury is a liquid at room temperature, freezing at −39° C. It boils at 357° C. Its vapor is poisonous. Like all metals it is an excellent conductor of electricity. Mercury is well below hydrogen in the activity series of metals and thus is not very active chemically, being attacked only by the oxidizing acids HNO_3 and concentrated H_2SO_4.

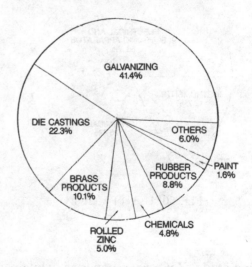

Fig. 49. Uses of Zinc

Great quantities of zinc are used in **galvanized iron.** The iron or steel is coated with zinc by dipping into molten zinc or by electrodeposition. Water pails, guttering, and wire fencing are among the many objects made of galvanized iron. Zinc forms the case and negative pole of dry cells, and is used in many types of electrical connectors. It is likewise an important constituent in many alloys. In brass and bronze it is alloyed with copper. Zinc **die casting** is an alloy with aluminum used in radio and automotive parts. Copper pennies contain 97.6% zinc and 2.4% copper.

Because cadmium is even less active than zinc, it is an excellent plating metal. It is present in several low-melting alloys used in fire sprinkling systems, and in wear-resistant alloys used as bearings.

Mercury is used to fill thermometers and barometers used in scientific work. The fact that it is a liquid makes it ideal for certain types of electrical switches. Mercury vapor lamps are commonly used in lighting, and the vapor is commonly used in neon signs. Alloys of mercury are called **amalgams.** Mercury is commonly used as a cathode in industrial electrolytic cells where it absorbs the metal being deposited as an amalgam and is later separated from it by distillation.

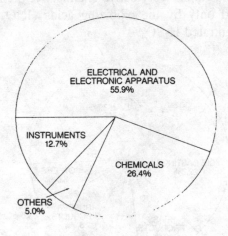

Fig. 50. Uses of Mercury

Zinc oxide, ZnO, is an important white paint pigment. It is used as a filler in the making of tire rubber, and its antiseptic properties are utilized in salves. A paste of zinc oxide and concentrated hydrochloric acid sets as a hard cement and is used in dentistry. **Zinc chloride,** $ZnCl_2$, is used as a wood preservative and as a flux in soldering and brazing. **Cadmium sulfide,** CdS, is a yellow paint pigment. **Mercurous chloride,** Hg_2Cl_2, is

known as **calomel** and is used as medicine to stimulate the secretion of bile by liver. **Mercuric chloride,** $HgCl_2$, is known as **corrosive sublimate,** and is a deadly poison. It is used as a germicide. **Mercury fulminate,** $Hg(ONC)_2$, explodes when struck. It is of great military importance as a primer in the firing of ammunition.

GALLIUM, INDIUM, AND THALLIUM

These three very rare metals have only recently been exploited. They are found as impurities in the sulfide ores of common metals, and are being marketed as by-products. Their low melting points make them valuable in low-melting alloys. **Gallium,** Ga, melts at 30° C. It will turn to liquid in the palm of one's hand. **Indium,** In, melts at 157° C., and **thallium,** Tl, melts at 303° C. Indium is likewise used as a protective coating for iron or steel, particularly in certain types of bearings.

Gallium arsenide, GaAs, is used in solar cells to convert sunlight into electricity. **Thallium oxide,** Tl_2O_3, has been used to make glass with a high index of refraction.

GERMANIUM, TIN, AND LEAD

Although none of these metals are very abundant, tin and lead occur concentrated in easily worked deposits, while germanium, which is at least as abundant as tin, is thinly distributed and is therefore less well known. **Germanium** is usually found associated with the sulfide ores of the common metals. The chief ore of **tin** is **cassiterite,** SnO_2, which is found in Malaya and Bolivia. **Galena,** PbS, is the chief source of **lead,** although this metal is also extracted from **cerussite,** $PbCO_3$; **anglesite,** $PbSO_4$; and **crocoite,** $PbCrO_4$.

Tin is easily reduced from its oxide ore by heating with carbon:

$$SnO_2 + 2\,C = 2\,CO + Sn.$$

If the process is carried out in a sloping furnace, the molten reduced metal flows from the furnace into molds.

Lead is reduced from its sulfide ore by a process of partial roasting followed by reduction. The sulfide is first heated to a relatively low temperature in air causing part of the sulfide to react as follows:

$$2 PbS + 3 O_2 = 2 PbO + 2 SO_2.$$

$$PbS + 2 O_2 = PbSO_4.$$

Then the temperature is raised to the **smelting** temperature where the remaining lead sulfide reacts with the products formed above as follows:

$$PbS + 2 PbO = 3 Pb + SO_2.$$

$$PbS + PbSO_4 = 2 Pb + 2 SO_2.$$

The molten lead runs from the smelter and is cast into molds.

Germanium may be extracted from the residues of refined lead or zinc electrolytically or by fractional crystallization.

Germanium is hard and brittle, and is a relatively poor conductor of electricity. However, its crystals have the property of being able to rectify alternating current, and for this reason crystals of germanium have been used as semiconductors in the electronics industry. Solar photovoltaic cells in the space program have also included germanium cells.

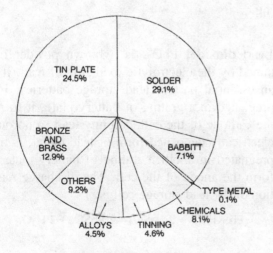

Fig. 51. Uses of Tin

Tin is a soft white low-melting metal. When exposed to cold weather for an extended time, it reverts to an allotropic form known as **tin disease.** It is only slightly above hydrogen in the activity series of metals and so is only moderately active chemically. It exhibits valence numbers of 2 and 4 in its compounds. It does not tarnish in air, being protected by a thin oxide film. It is used chiefly as **tinplate** in the protection of iron and steel, and in many important nonferrous alloys (Table XLIII).

Lead is soft, nonelastic, low-melting, bluish-white metal that becomes dark gray in air from

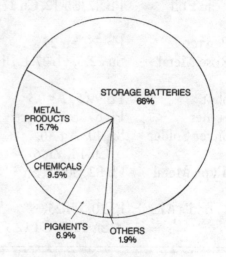

Fig. 52. Uses of Lead

an oxide film. This film then protects the metal from further oxidation. Lead is just above hydrogen in the activity series and is not seriously attacked by acids at ordinary temperatures. Water containing dissolved oxygen slowly oxidizes and dissolves lead, thus making it unfit for drinking-water piping, for all lead salts are poisonous and lead accumulates in the body. Lead, like tin, exhibits valence numbers of 2 and 4 in its compounds. Lead is used extensively for piping and as a coating around telephone and other electric cables to protect them from corrosion. It is a vital alloy metal. Table XLIII describes some of the important lead and tin alloys.

Stannic chloride, $SnCl_4$, is an important in-

Table XLIII
Alloys of Lead and Tin

Alloy	Composition	Properties	Uses
Babbitt Metal	Sn 89, Sb 7.3, Cu 3.7	Soft, low friction	Bearings
Battery Plate	Pb 94, Sb 6	Easily cast	Batteries
Bearing Metal	Sn 75, Sb 12.5, Cu 12.5	Soft, low friction	Bearings
Bell Metal	Cu 75, Sn 25	Clear tone when struck	Bells
Bronze	Cu 88, Sn 10, Zn 2	Corrosion resistant, easily cast	Bearings, castings, statues
Fuse Metal	Bi 50, Cd 10, Pb 26.6, Sn 13.3	Low melting	Fuses, safety fire sprinklers
Lead Foil	Pb 87, Sn 12, Cu 1	Malleable, inert	Wrappings for cigarettes, etc.
Pewter	Pb 75, Sn 25	Soft, easily worked	Ornamental ware
Rose Metal	Sn 22.9, Pb 27.1, Bi 50	Low melting	Safety fire sprinklers
Shot	Pb 99.5, As 0.5	Hard	Shotgun shot
Solder	Pb 67, Sn 33	Low melting	Soldering
Silver Solder	Sn 40, Ag 40, Cu 14, Zn 6	Low melting	Soldering
Type Metal	Pb 82, Sb 15, Sn 3	Low melting, expands on cooling	Printing type, castings
Wood's Metal	Bi 50, Pb 25, Sn 12.5, Cd 12.5	Low melting	Fuses, safety fire sprinklers

termediary in the recovery of tin from tin cans and other tin-plated objects. Chlorine gas is passed over the tin, forming the liquid chloride:

$$Sn + 2\ Cl_2 = SnCl_4.$$

This is then dissolved in water and boiled, causing stannic oxide to precipitate:

$$SnCl_4 + 2\ H_2O = \underline{SnO_2} + 4\ HCl.$$

Metallic tin is then reduced from the oxide by heating with carbon.

Litharge, PbO, **red lead,** Pb_3O_4, and **white lead,** $Pb_3(OH)_2(CO_3)_2$, are all important paint compounds. They have been used in the past and are still used in oil-based paints. However, with the development of water-based paints, the use of lead compounds in paints has been decreasing.

Lead dioxide, PbO_2, is a brown powder obtained by the electrolytic oxidation of lead. It is an essential part of lead storage batteries. Impregnated in a grating of battery plate, it forms the cathode of the cell. **Spongy lead,** a porous, electrically deposited form of lead metal, is impregnated in another grating of battery plate to form the anode of the cell. The discharge reaction of the lead storage cell is:

$$Pb + PbO_2 + 2\ H_2SO_4 = 2\ PbSO_4 + 2\ H_2O.$$

Heat shortens the life of a storage battery by causing the lead dioxide to decompose:

$$2 \, PbO_2 = 2 \, PbO + O_2.$$

Tetraethyl lead, $(C_2H_5)_4Pb$, is added in small quantities to gasoline to prevent engine knock. Its use is constantly decreasing, due to the advent of unleaded gasoline, which does not discharge toxic lead into the environment.

Problem Set No. 22

1. What is the green deposit that forms in a sink from a dripping faucet, and what is the origin of its constituents?
2. How does the electrolytic refining of copper using copper electrodes take full advantage of the electrochemical properties of metals and their ions?
3. Write equations for the reduction of zinc from sphalerite.
4. Give one application of each of the following metals which would make it critical in a national emergency. (a) Copper. (b) Tin. (c) Mercury.
5. Why is lead unfit for use as pipes for drinking water?

SUMMARY

Copper is second to iron in importance among the metals. It is fairly abundant, occurring free and in several important ores. Its sulfide ores are concentrated, roasted, and then reduced. The impure copper is refined electrolytically. Copper is used as an electrical conductor and in piping. Its principal alloys are **brasses, bronzes,** and **coinage metals.**

Zinc and **mercury** are likewise obtained from sulfide ores. Much zinc is used in **galvanized iron** and in alloys, particularly **brass** and zinc **die casting.** Mercury is a **liquid,** and is used in thermometers, barometers, and electrical switches.

Tin is reduced by carbon from its oxide ore. It is used primarily as **tinplate** and in **low-melting** and **bearing alloys.**

Lead is obtained from its sulfide ore by partial roasting followed by **smelting.** It is used in piping, in batteries, and in low-melting alloys. **Red lead** and **white lead** are important ingredients in **paint.**

CHAPTER 21

THE NOBLE AND RARE METALS

Silver, gold, and platinum are called the **noble metals** because they occur free in nature. However, the term may likewise be used with reference to any of the other metals of low chemical activity which are found in the free state. A metal found free is also referred to as a **native metal.**

SILVER AND GOLD

The use of these two metals as coinage and for ornamental ware has caused them to be associated with wealth and the finer things of life. As a matter of fact, iron, copper, and other common metals are of much more economic value than these, and some of the rare metals cost considerably more per pound.

Silver is found free and as the sulfide **argentite,** Ag_2S, usually associated with copper and lead sulfide ores. Most of the metallic silver produced is obtained from the anode slimes from the electrolytic refining of copper and lead. Silver is also obtained from lead produced by the smelting process. Zinc is stirred into molten lead, but the two molten metals are immiscible. Silver dissolves in the zinc, which floats on the lead and is skimmed off. The zinc is then distilled away from the silver. This is known as the **Parkes process.** Silver is extracted from silver ore by treating the crushed ore with sodium cyanide, which dissolves the silver sulfide. Metallic zinc is then added to reduce the silver from solution, the reaction being a direct application of the activity series of metals. Silver is refined electrolytically.

Gold is usually found free as **nuggets** in quartz rock formations or as grains or flakes in sands formed by the weathering of siliceous minerals. Gold-bearing rock is crushed and treated with sodium cyanide, which dissolves the gold just as in the case of silver. Again zinc is added to cause the less active gold to precipitate from solution. The high density of gold (19.3) is utilized in **panning** and **sluicing** operations in which water washes away the less dense rock, leaving the heavy gold particles behind. Mercury is frequently used to absorb gold from sand as an amalgam. The amalgam is then heated to drive off the mercury. Gold is refined electrolytically.

Silver is a bright, lustrous metal which is soft and very malleable. It is easily worked into complex shapes, and hence is much used for jewelry and tableware. It is extensively used as a plating metal, being deposited electrolytically from cyanide solutions. Its ability to reflect light is utilized in mirrors. It is not corroded by the atmospheric gases, but sulfur compounds tarnish it badly forming black silver sulfide, Ag_2S. Tarnished silver objects may be cleaned by boiling them in water containing a spoonful of baking soda in an aluminum pan. An electrochemical reaction takes place in which aluminum dissolves as ions and silver ions in the sulfide are redeposited on the surface of the object. The baking soda serves as an inert electrolyte. The overall reaction is:

$$3\ AgS + 2\ Al + 6\ H_2O = 2\ Al(OH)_3 + 3\ H_2S + 3\ Ag.$$

Buffing the object then restores the luster of the silver.

EXPERIMENT 29: Place a tarnished silver spoon in a clean aluminum pot and cover it with water to a depth of 1–2 inches. Add a teaspoon of baking soda. Boil for several minutes. Remove the silver spoon and buff it with a soft cloth.

Silver is attacked by nitric and concentrated sulfuric acids, but it is inert to alkalis.

Pure silver is too soft for use as coinage and tableware. Consequently a little copper is alloyed with it to harden and strengthen it. **Sterling silver** contains 7½% copper. The amount of silver in an alloy is often indicated as **fineness,** which is the parts by weight of silver per 1000 parts of alloy. Thus, U.S. coinage is 900 **fine.**

The yellow color and the luster of gold are well known. It is a soft metal and is extremely ductile and malleable. It can be drawn into extremely fine thread, and is beaten between sheets of parchment paper into **gold leaf** as thin as 0.00002 mm. thick. It is very inert to ordinary reagents, although chlorine gas will corrode it. It dissolves in **aqua regia,** forming a solution of **chlorauric acid,** $HAuCl_4$.

Gold leaf, used in window signs, book binding, electrical instruments, and for various decorative purposes, is the chief application of pure gold. The gold used in coinage, jewelry, and dentistry is alloyed with other metals to give it hardness. Copper, nickel, and zinc are usually used. The gold content of these alloys is rated in **carats,** indicating the parts by weight of gold in 24 parts by weight of alloy.

PHOTOGRAPHY

Silver halides are sensitive to light energy. The precise nature of this sensitivity has not yet been fully explained, but when light strikes the silver ions in these salts they become more susceptible to chemical reduction. This phenomenon is utilized in photographic film and paper.

Photographic film consists of either a thin film of polyester plastic called **Mylar** or **Cronar,** or, to a lesser extent, of a thin film of plastic called **cellulose acetate,** on which there is a thin film of silver bromide and silver iodide suspended in gelatin.

There are four basic steps involved in obtaining a picture from photographic film: exposure, developing, fixing, and printing.

During **exposure,** light is permitted to strike the sensitive layer on photographic film. The silver ions struck by the light are made susceptible to easy reduction to metallic silver in direct proportion to the intensity of the light striking the ions. It should be pointed out that examination of the silver salts after brief exposure does not reveal any evidence of either chemical or physical change in the ions, although prolonged exposure will cause a reduction to fine specks of jet black metallic silver known as **colloidal silver.**

In the **developing** process, the exposed film is immersed in a solution of a mild reducing agent

Table XLIV
Alloys of Noble Metals

Alloy	*Composition*	*Properties*	*Uses*
Sterling silver	Ag 2.5, Cu 7.5	Corrosion resistant, easily worked	Coinage, jewelry, tableware
U.S. commemorative silver coins	Ag 90, Cu 10	Corrosion resistant, durable	Coinage
U.S. commemorative gold coins	Au 90, Cu 10	Corrosion resistant, durable	Coinage
18-carat yellow gold	Au 75, Ag 12.5, Cu 12.5	Corrosion resistant, easily worked	Jewelry, accessories
18-carat white gold	Au 75, Cu 3.5, Ni 16.5, Zn 5	Corrosion resistant, easily worked	Jewelry, accessories
Inlay	Au 90, Ag 1.5, Cu 7, Zn 0.5, Pt 1.0	Corrosion resistant, easily worked	Decoration
14-carat dental gold	50 Au, 33 Ag, 17 Cu	Corrosion resistant, easily worked	Dental fillings and braces

such as pyrogallol, $C_6H_3(OH)_3$, or hydroquinone, $C_6H_4(OH)_2$. These reducing agents cause a reduction of only the silver ions made sensitive by light during exposure. The developing solution also contains Na_2CO_3, sodium carbonate, because the developing action takes place only in mildly alkaline solution, and sodium sulfite, Na_2SO_3, which prevents attack upon the developers by atmospheric oxygen. The developed film is thus coated with black colloidal silver wherever white light originally struck it. Hence it is called a **negative.** When the developing is completed, the film is dipped into a solution of acetic acid which neutralizes the sodium carbonate and thus stops the reducing action of the developers.

In the **fixing** process, the unexposed silver salts are removed from the film. The fixing solution contains sodium thiosulfate, $Na_2S_2O_3$. This is known as **hypo.** It reacts with the silver salts still in the gelatin as follows.

$$2\,AgBr + 3\,Na_2S_2O_3 = 2\,NaBr + Na_4Ag_2(S_2O_3)_3.$$

Both the sodium bromide and the sodium silver thiosulfate formed in this reaction are soluble, and thus the fixing solution washes the silver salts from the film. The fixing solution also contains a little acetic acid and some alum which serves to harden the gelatin film. When the fixing is completed, the film is thoroughly washed with water to remove the excess hypo which would otherwise bleach the negative. The negative is then dried.

In the **printing** process, the developed negative is held against a sensitized film on the printing paper. The film on this paper usually contains silver chloride suspended in gelatin. Silver chloride is less sensitive to light than the other silver halides. Light is then permitted to pass through the negative to strike the silver ions on the printing paper. This exposes them just as the original film was exposed, except that this time the light strikes most heavily on areas that were dark when the original picture was taken. The black silver on the negative shields the bright areas in the original object. The exposed printing paper is then developed and fixed in the same manner as the film. The finished print is thus a reproduction of the original object and is known as a **positive.**

The film used in color photography has three sensitized layers each containing a dye which is absorbed by metallic silver in the layer to produce the three primary colors, red, yellow, and blue, when the film is developed. When white light is passed through the developed film, not only are the light and dark areas of the object reproduced, but also all the hues and shades of its colors. The printing of colored film involves the use of special paper with several layers of silver salts containing dyes capable of reflecting the colors of the original object when the paper is developed.

THE PLATINUM METALS

The six metals below iron, cobalt, and nickel in the periodic table are known as the platinum metals. The first three, ruthenium, Ru, rhodium, Rh, and palladium, Pd, are the light platinum metals, and the last three, osmium, Os, iridium, Ir, and platinum, Pt, are the heavy platinum metals. They are all very inert except toward alkalis, and all occur free in nature. They are quite rare and very expensive. Osmium has the greatest density of all the elements weighing 22.5 grams per cc. None of them have important compounds.

Platinum is the most important of these metals, being used especially in jewelry because it does not tarnish and is easily worked. It is likewise used in dentistry and in the construction of electrical apparatus where its excellent conductivity is utilized. Finely divided platinum is an important industrial catalyst, especially in the contact process of making sulfuric acid. The international standards of weights and measures are made of platinum-iridium alloys.

THE INNER TRANSITION METALS

The last remaining metals of the periodic table include **scandium,** Sc, **yttrium,** Y, and the 30 metals of the **Lanthanide** and **Actinide series,** or **inner transition series,** of metals. These metals are rare, with uranium being the most abundant. The Lanthanide metals were originally called the rare earth metals. The fact that the inner transition series metals differ structurally in the third outermost shell of electrons makes them extremely difficult to separate and isolate. Thus, they have few uses. **Uranium,** U, **thorium,** Th, and **plutonium,** Pu, have isotopes which are sources of atomic energy, a matter to be discussed in Chapter 23. An alloy of **cerium,** Ce, and iron, known as Auer metal, gives off sparks when struck, and is consequently used in pocket lighters as "flints" and as tracer lights in tracer bullets. A mixture of cerium and thorium oxides forms the wick of gas mantel lamps. **Cerium sulfate,** $Ce(SO_4)_2$, is used in analytical chemistry. **Europium oxide,** Eu_2O_3, is a brilliant red phosphor used in TV receivers. The oxides of **lanthanum, praseodymium,** and **neodymium** (La_2O_3, Pr_2O_3, Nd_2O_3) are used in sunglasses to absorb injurious uv radiation and to reduce the sunlight. **Uranium hexachloride,** UF_6, was the gaseous compound used to separate the uranium isotopes in making the original atomic bomb.

Problem Set No. 23

1. Would a noble metal serve as an effective sacrificial anode (See Chapter 11)?
2. What is aqua regia?
3. To what do the terms "fineness" and "carats" refer?
4. Could you suggest a method of recovering silver from used fixing solutions?
5. Using the table of atomic weights in this book, list the names of the lanthanide and actinide series of metals.

SUMMARY

Noble metals, particularly silver, gold, and platinum metals, are very inert and occur free in nature.

Silver is extracted from its ores by the cyanide process, but most silver is obtained as a by-product from the electrolytic refining of other metals. The **Parkes process** is a method of obtaining silver from lead in the impure state.

Gold is found free, and is also separated from rock by the cyanide process.

Silverware may be **cleaned** by boiling it in a solution of baking soda in an aluminum pan.

The sensitivity of silver halides to light makes them useful in photography. Color photography combines dyes with the sensitive silver salts on the film and printing paper to produce the colored reproductions of objects.

Platinum is the most important of the six platinum metals, and is extensively used in jewelry.

The **inner transition metals** are not only rare, but their structural similarity makes it difficult to isolate them.

CHAPTER 22

ORGANIC CHEMISTRY

Early scientists discovered that the compounds of carbon showed a different chemical behavior than the compounds of the other elements. They react more slowly, in general, and have a tendency to form equilibrium mixtures rather than go to completion in their reactions. The conditions of temperature and pressure affect reactions of these substances much more than in the case of the compounds of other elements, and it was soon discovered that it frequently happens that carbon compounds of entirely different properties have the same percentage composition. Since these compounds were originally obtained from living things or from the remains of living things, like coal, the study of these compounds was called **organic chemistry.** It was later discovered that organic compounds could indeed be formed from wholly inorganic raw materials.

The field of organic chemistry is so vast that, despite the fact that thousands of organic compounds are in daily use, only a tiny fraction of the possible compounds are being utilized. Yet organic chemicals play a vital role in the life of every person in food and beverages, in drugs and medicines, in textiles and dyes, in plastics and roads, as fuels or refrigerants, as explosives or adhesives.

The key element present in all organic chemicals is carbon. It has a valence of 4, for it has 4 electrons in its outermost shell which are used in forming covalent bonds with atoms of hydrogen, oxygen, the halogens, nitrogen, sulfur, and other carbon atoms. Unlike the situation in inorganic chemistry where electrovalent, ionic compounds predominate, the properties and chemistry of carbon compounds are intimately related to the nonionic, covalent bonds present in these compounds. Single, double, and even triple covalent bonds between a single pair of atoms, especially carbon atoms, are encountered. More specifically, the properties and chemistry of carbon compounds are related to three factors:

1. The number of carbon atoms in the molecule.

2. The type of bonding (single, double, or triple) present.
3. The kinds of **functional groups** present in the molecule.

Simple organic molecules have been isolated which contain 60 or more carbon atoms, and complex molecules consisting of endless chains of structural units, like starch or rubber, contain staggering numbers of carbon atoms. In general, an increase in number of carbon atoms tends to increase the melting and boiling point of the compounds and to reduce the chemical activity of them.

A single covalent bond consists of a single pair of electrons shared between two atoms. Double bonds involve two shared pairs of electrons, and triple bonds involve three shared pairs. The bonding in organic compounds is usually indicated by dashes as in the following examples:

Methane–CH_4 Ethylene–C_2H_4 Acetylene–C_2H_2

In general, the single bond is the least active chemically. The double bond is more reactive, and compounds containing double bonds are somewhat more volatile than corresponding singly bonded compounds. The triple bond is the most active chemically, and the volatility is still greater in compounds containing this type of bond.

The functional groups of organic chemistry are similar to the radicals of inorganic chemistry, in that they behave as a unit in entering or leaving an organic molecule and infuse particular properties in molecules possessing them. Table XLV gives the names and formulas of the principal functional groups.

The dashes in the formulas of the functional groups in Table XLV indicate the location of the

Table XLV
Organic Functional Groups

Name	Formula	Name	Formula
Methyl	$-CH_3$	Alcohol, or Hydroxy	$-OH$
Ethyl	$-C_2H_5$	Aldehyde	$-C{\big<}^O_H$
Propyl	$-C_3H_7$	Ketone	$\overset{O}{\overset{\|}{-C-}}$
Butyl	$-C_4H_9$	Acid	$-C{\big<}^O_{OH}$
Phenyl	$-C_6H_5$	Ether	$-O-$
Chloro	$-Cl$	Ester	$-C{\big<}^O_{O-}$
Bromo	$-Br$	Nitrile	$-C{\equiv}N$
Iodo	$-I$	Amino	$-NH_2$
Fluoro	$-F$	Nitro	$-NO_2$

bonds available for attachment to basic organic molecules, or to other groups.

Because the chemistry of organic compounds is so closely related to the functional groups present in the molecules, the usual empirical formulas of compounds showing only the ratio of the number of each type of atom present in the molecule is not sufficient to indicate just which compound is under consideration. Instead, **structural formulas** are used. The formulas of organic compounds must indicate in some manner the structure of the molecles as well as their composition. Consider ethyl alcohol. It is composed of an ethyl group, $-C_2H_5$, combined with an alcohol group, $-OH$. Its *complete structural* formula is indicated as follows:

$$H-\overset{\overset{\textstyle H}{|}}{\underset{\underset{\textstyle H}{|}}{C}}-\overset{\overset{\textstyle H}{|}}{\underset{\underset{\textstyle H}{|}}{C}}-O-H \text{ (ethyl alcohol)}$$

Frequently, in order to conserve space and time, a modified formula known as a **condensed structural formula** is used to represent an or-

ganic compound. In the case of ethyl alcohol, the condensed structural formula would be C_2H_5OH. Note that this formula indicates clearly the functional groups present, but it is less effective in showing exactly where the bonds are located, especially the carbon-oxygen bond observed in the complete structural formula above.

The empirical formula of ethyl alcohol would be C_2H_6O. This in no way indicates the functional groups present. Furthermore, there are many instances in organic chemistry when different compounds possess identical empirical formulas. For example, CH_3OCH_3, dimethyl ether, has the same empirical formula as ethyl alcohol, but it has totally different properties because it possesses different functional groups. Different compounds possessing the same empirical formulas are known as **isomers.**

There are two basic types of organic reactions, depending upon the bonding involved.

1. **Substitution reactions.** This type of reaction usually involves the single covalent bond and it results in the replacement of one atom or functional group by another. For example, if the gas

methane, CH_4, is treated with chlorine gas in the presence of ultraviolet light, the following series of reactions take place, each involving the substitution of a chlorine atom for a hydrogen atom:

$CH_4 + Cl_2 = HCl + CH_3Cl$ (Methyl chloride)
$CH_3Cl + Cl_2 = HCl + CH_2Cl_2$ (Methylene chloride)
$CH_2Cl_2 + Cl_2 = HCl + CHCl_3$ (Chloroform)
$CHCl_3 + Cl_2 = HCl + CCl_4$ (Carbon tetrachloride)

It should be pointed out that each of the products from this series of reactions is commercially important: methyl chloride as an anesthetic, methylene chloride as a solvent, chloroform as a solvent and an anesthetic, and carbon tetrachloride as a solvent.

A substituted methane, **halon,** $CClBrF_2$, is now widely used in portable fire extinguishers. It is a vaporizing liquid that can put out fires with much lower concentrations than carbon dioxide, CO_2.

2. **Addition reactions.** This type of reaction involves only double or triple bonds, resulting in the splitting of one of the bonds and the addition of a functional group to each of the original atoms in the multiple bond. For example, ethylene, $H_2C = CH_2$, adds chlorine gas readily with one chlorine atom going to each of the carbon atoms:

$H_2C = CH_2 + Cl_2 = CH_2ClCH_2Cl.$ (Dichloroethane).

The compounds of organic chemistry may be divided into two general classes, aromatic and aliphatic. **Aromatic compounds** all contain a molecular structure consisting of one or more benzene rings. Benzene, C_6H_6, is the compound of this class and has its six carbon atoms arranged in a hexagon with alternate single and double bonds between them. One hydrogen atom is bonded to each carbon atom. The structure may be represented:

Benzene

Other functional groups may then replace the hy-

drogen atoms to form an almost endless variety of aromatic compounds.

Aliphatic compounds include all organic compounds which do not possess a benzene ring. Their basic structure involves chains of carbon atoms which may be either straight, but puckered or coiled, or branched. The carbon atoms may also be arranged in rings which are different from the benzene ring. Such compounds are called **alicyclic** compounds. Again, an almost endless variety of compounds of this class are possible.

From the point of view of composition, the most common organic compounds fall into the following classes:

1. Hydrocarbons.
2. Oxidation products of hydrocarbons including alcohols, aldehydes, ketones, and acids.
3. Ethers.
4. Esters.
5. Derivatives, which are basically hydrocarbons which have one or more hydrogen atoms replaced by functional groups.
6. Carbohydrates.

The basic organic compounds which serve as raw materials for the preparation of the many specific synthetic compounds are obtained from three sources:

1. **Coal tar.** When coal is heated in the absence of air (destructive distillation) in huge ovens to form coke, all of the volatile matter escapes from the coal. It is collected and condensed to a black tarry liquid called coal tar. This is the chief source of aromatic hydrocarbons.

2. **Petroleum** and **natural gas.** Petroleum is a mixture of principally aliphatic hydrocarbons. Many of these are liquid at ordinary temperatures, and these in turn serve as solvents for low molecular weight compounds which would otherwise be gases, and high molecular weight compounds which would otherwise be solids. Natural gas consists of the very light hydrocarbons that have escaped from the petroleum solution. Petroleum is thus the chief source of aliphatic hydrocarbons.

3. **Plant life.** Plants form the principal source of the carbohydrates: sugar, starch, and cellu-

Table XLVI
Homologous Series

Name	*Formula*	*Familiar Members*
Methane, Alkane, or Paraffin	C_nH_{2n+2}	Methane—CH_4; ethane—C_2H_6; Propane—C_3H_8; butane—C_4H_{10}; Octane—C_8H_{18}.
Ethylene or Alkene	C_nH_{2n}	Ethylene—C_2H_4.
Acetylene or Alkyne	C_nH_{2n-2}	Acetylene—C_2H_2.
Cycloparaffin or Naphthenes	C_nH_{2n}	Cyclopentane—C_5H_{10}.
Benzene	C_nH_{2n-6}	Benzene—C_6H_6; toluene—$C_6H_5CH_3$.

lose, as well as some special hydrocarbons such as turpentine and natural rubber.

HYDROCARBONS

Hydrocarbons are compounds of carbon and hydrogen. Several important families of hydrocarbons, each containing many compounds, are well known. Such families of hydrocarbons are known as **homologous series,** and all the members have formulas that fit the general formula of the series. The principal homologous series of hydrocarbons are summarized in Table XLVI.

Although the alkene and cycloparaffin series have the same formula, they behave chemically quite differently. In the alkene series there is a double bond present between two carbon atoms which greatly increases the chemical reactivity of these hydrocarbons. In the cycloparaffin series, the carbon atoms are joined to each other, forming a ring. Such a structure is relatively inert.

Both the paraffin and the cycloparaffin series are said to be **saturated** because they contain only single bonds. The other series are all **unsaturated** because they contain either double or triple bonds. Because of their widespread and important uses, some of the hydrocarbons deserve special mention.

Methane, CH_4. This gaseous compound is the chief ingredient of natural gas. Under the name **marsh gas** it escapes from peat bogs and is responsible for the eerie will-o'-the-wisp observed when it becomes ignited over these swamps. Under the name **fire damp** it is the gas that escapes from coal, which has been responsible for many disastrous mine explosions. It is highly combustible and is an excellent fuel. Like all of the other basic hydrocarbons, it is an important raw material for the manufacture of synthetic organic chemicals.

Ethane, C_2H_6. This gas makes up about 10% of natural gas. Its uses are similar to those of methane.

Propane, C_3H_8. This is one of the commonly used ingredients in bottled gas. It is also used as a high-pressure solvent

Butane, C_4H_{10}. In addition to being useful as a bottled gas and as a solvent, butane is used in the production of high-test gasoline. Actually there are two butanes, **normal butane,** containing a straight chain of carbon atoms, and **isobutane,** which has a branched carbon chain. It is this isomer that is used in making special gasoline. The prefix "normal" (n-) always refers to a straight chain of carbon atoms. The prefix "iso-" always refers to a branched chain of carbon atoms. The structures of the isomers of butane are as follows:

n-Butane   iso-butane

Octane, C_8H_{18}. This compound has 18 isomers, one of which, $(CH_3)_3CCH_2CH(CH_3)_2$, isooctane, is the reference compound for rating the **octane number** of gasoline. By definition, the octane number of a gasoline is the percentage of isooctane that must be added to n-heptane, C_7H_{16}, to cause a standard engine to operate with the same characteristics as the gasoline being tested. Pure isooctane has a rating of 100. High-test automotive gasolines now have octane numbers between 90 and 98. Aviation gasolines have octane numbers above 100, which simply means that they give better performance in the standard engine than pure isooctane.

Petroleum refining. In the refining of petroleum, many useful mixtures of hydrocarbons are made available. The basic process in refining is **fractional distillation,** which separates the various products according to ranges of boiling points. The products formed by the initial distillation of crude petroleum are known as **straight run** products. Table XLVII gives the main products formed, together with other pertinent information about them.

Since gasoline is only a small portion of the straight run product, and since the demand for it is so great, other steps are introduced into the refining process to increase the yield of gasoline. The two basic steps are:

1. **Polymerization.** In this process, molecules containing less than the proper number of atoms of carbon for gasoline are caused to be linked together to form larger molecules which fit into

Table XLVII
Petroleum Products

Name	Boiling Range °C.	No. of Carbon Atoms Present
Gas	Below 32	1–4
Gasoline	40–200	4–12
Naphthas	50–200	7–12
Kerosene	175–275	12–15
Fuel Oil	200–300	15–18
Lubricating Oil	Above 300	16–20
Wax	Above 300	20–34
Asphalt	Residue	Large

the gasoline range under the influence of proper catalysts.

2. **Cracking.** In this process, the large molecules from the products beyond the gasoline range are broken apart, either by heat or by catalytic action, to fit the gasoline range.

Refined gasoline, especially high-test gasoline, may contain additives to assist in its proper combustion. In 1922 $(C_2H_5)_4Pb$, tetraethyl lead, was first added to gasoline to prevent **knocking,** a condition created when the gasoline is ignited too soon in the engine. Since the combustion of tetraethyl lead causes deposits of metallic lead to form on spark plugs and in the cylinders, ethylene dibromide, $C_2H_4Br_2$, was then added to remove the lead as volatile lead bromide, $PbBr_2$.

Since 1975, cars made in the United States have been installed with catalytic converters in the exhaust system. These converters help reduce air pollution. However, they are fouled by lead compounds. No-lead gasoline, which has a high octane number, has been developed by replacing tetraethyl lead with additives such as benzene and toluene, and by increasing the ratio of branched-chain to straight-chain hydrocarbons in the gasoline.

Rubber, $(C_5H_8)_n$. Natural rubber consists of tremendously long chains of polymerized C_5H_8 units. The long molecules are coiled and are easily stretched. When rubber is **vulcanized,** sulfur atoms become bonded between carbon atoms of different chains, resulting in better wearing properties over a wider temperature range. The addition of fillers like zinc oxide and carbon black makes the rubber less susceptible to oxidation.

Benzene, C_6H_6. This liquid is the fundamental hydrocarbon of the aromatic class. It is one of the principal ingredients of coal tar and is refined from it by fractional distillation. It is an important solvent and a fundamental raw material from which a multitude of other chemicals are made.

Toluene, $C_6H_5CH_3$. This compound may also be called methylbenzene. It is obtained along with benzene from coal tar and is used both as a solvent and as a raw material for making other chemicals, including explosives like trinitrotoluene, TNT, $C_6H_2CH_3(NO_2)_3$.

Naphthalene, $C_{10}H_8$. This white crystalline solid has long been used as mothballs. It is obtained from coal tar and is likewise a raw material for preparing other chemicals. It has a structure consisting of two benzene rings joined side by side. Paradichlorobenzene, $C_6H_4Cl_2$, is also used for mothballs.

Anthracene, $C_{14}H_{10}$. This crystalline fluorescent solid is also obtained from coal tar, and it is used in the preparation of dyes. Its structure consists of three benzene rings joined side by side.

OXIDATION PRODUCTS

Hydrocarbons are highly combustible substances which burn when vigorously oxidized to form CO_2 and H_2O. However, if the oxidation of hydrocarbons is carried out gently, a number of important intermediate compounds can be formed. The mild oxidation of hydrocarbons involves either the insertion of an oxygen atom into the molecule, or the removal of two hydrogen atoms from the molecule. We may summarize the mild oxidation of methane as follows:

$$CH_4 \xrightarrow{(+O)} CH_3OH \xrightarrow{(-H_2)} HC\!\!\begin{array}{c} \nwarrow O \\ \diagup H \end{array} \xrightarrow{(+O)}$$

| Methane | Methyl Alcohol | Formaldehyde |

$$HC\!\!\begin{array}{c} \nwarrow O \\ \diagup OH \end{array} \xrightarrow{(+O)} CO_2 + H_2O.$$

| Formic Acid | End Products |

Thus between the hydrocarbon and the end products of oxidation three types of compounds appear: alcohols, aldehydes, and organic acids.

It is sometimes necessary in actual practice to carry out the steps in the oxidation of hydrocarbons indirectly. For example, in making methyl alcohol from methane, the methane may be first treated with chlorine to form methyl chloride, which is then caused to react with sodium hydroxide to produce the alcohol according to the following reactions:

$$CH_4 + Cl_2 = HCl + CH_3Cl$$
$$CH_3Cl + NaOH = NaCl + CH_3OH.$$

The oxidation of an alcohol like isopropyl alcohol, $CH_3CHOHCH_3$, where the alcohol group is in a nonterminal position in the carbon chain, produces not an aldehyde but a compound known as a **ketone.**

$$\begin{array}{c} CH_3 \\ \diagdown \\ CH-OH \\ \diagup \\ CH_3 \end{array} \xrightarrow{(-H_2)} \begin{array}{c} CH_3 \\ \diagdown \\ C=O. \\ \diagup \\ CH_3 \end{array}$$

| Isopropyl Alcohol | Acetone |

Ketones cannot be oxidized to acids, but vigorous oxidation (burning) again yields CO_2 and H_2O.

Organic acids such as CH_3COOH, acetic acid, are weak electrolytes in general. They form salts with bases just as inorganic acids do.

ALCOHOLS

Methyl alcohol, CH_3OH. This volatile liquid is also known as methanol and as wood alcohol. The latter name is based on the fact that this compound is an important by-product from the destructive distillation of wood to produce charcoal. Great quantities of it are also produced by the catalytic hydrogenation of carbon monoxide. This compound is poisonous and can produce blindness. Since it is absorbed through the skin, it is unfit for use as rubbing alcohol. It is a very important solvent and is used both as an antifreeze and as a denaturant for ethyl alcohol. It is the starting point in the production of many synthetic chemicals.

Ethyl alcohol, C_2H_5OH. This vitally important compound is produced in large quantities from the fermentation of carbohydrate compounds contained in molasses, corn, rye, barley, and potatoes. Enzymes in yeast cause the fermentation. It has the property of absorbing moisture from the atmosphere until it has a composition of 95% alcohol and 5% water. It is also known as grain alcohol and as ethanol. It is extensively used in the production of other chemicals and chemical products, particularly in industries preparing drugs, medicinals, and cosmetics. It is present in a variety of beverages, causing them to be intoxicating. It ranks next to water as an important solvent, and alcoholic solutions are known as tinctures. To prevent its illegal use in the manufacture of beverages, it is often denatured by dissolving poisonous and nauseating compounds in it.

Isopropyl alcohol, $CH_3CHOHCH_3$. This liquid is extensively used as rubbing alcohol. It is prepared from the gases obtained from the cracking of petroleum.

Glycerin, $C_3H_5(OH)_3$. This sweet, syrupy liquid is obtained as a by-product from the manufacture of soap. It is used as an antifreeze and as a moistening agent in tobacco and cosmetics. When treated with concentrated nitric and sulfuric acids it is converted to nitroglycerin, which is in turn absorbed in clay to form dynamite.

Phenol, C_6H_5OH. Known also as carbolic acid, this compound is obtained in great quantity from coal tar. It is an important disinfectant, and great quantities of it are used in the preparation of plastics and other aromatic chemicals. It may also be called monohydroxybenzene. Its related polyhydroxybenzenes such as **hydroquinone,** $C_6H_4(OH)_2$, and **pyrogallol,** $C_6H_3(OH)_3$, are important photographic developers.

ALDEHYDES

Formaldehyde, HCHO. This gas is prepared by oxidizing methyl alcohol with hot copper oxide. A 40% solution of it is called formalin. It is poisonous, and is an important germicidal agent and preservative. Great quantities of this compound are used with phenol in the preparation of plastics.

Acetaldehyde, CH_3CHO. Small amounts of this and other aldehydes are responsible for the flavor and scent of alcoholic beverages. It is formed by the gradual oxidation of ethyl alcohol. It is used as an intermediate in the preparation of other organic chemicals because it is very reactive chemically.

KETONES

Acetone, CH_3COCH_3. This volatile liquid is formed by the oxidation or dehydrogenation of isopropyl alcohol. It is also obtained from the destructive distillation of wood. It is an important solvent as well as a raw material in the preparation of other organic compounds.

ACIDS

Acetic acid, CH_3COOH. Some acetic acid is obtained from the destructive distillation of wood, but most of it is produced by the fermentation of fruit juices. **Vinegar** is a dilute solution of this acid. Pure acetic acid is known as **glacial** acetic acid. It is an excellent solvent, and is used in the preparation of photographic film and synthetic fibers.

Formic acid, HCOOH. Produced by the oxidation of formaldehyde, this liquid with an irritating odor is found in ants (from which it was originally prepared by distillation) and stinging plants. It is a constituent of perspiration.

Citric acid, $H_3C_6H_5O_7$. This solid compound

is contained in the juices of citrus fruits like lemons and limes, and was formerly obtained almost exclusively from these fruits. It is now prepared by the action of molds on molasses and sugar. It is used as an ingredient in many drugs and in soft drinks.

Oxalic acid, $H_2C_2O_4$. This solid acid is produced by the action of the same molds on molasses and sugar as in the production of citric acid. When the pH of the solution is 6–7, oxalic acid is formed. When the pH is reduced to 1–2, citric acid is produced. Oxalic acid is an important reducing agent and is used to bleach wood and rust stains in clothing.

ESTERS

Esters are compounds formed from the reaction of an alcohol with an organic acid. The general reaction is:

$$\text{Alcohol} + \text{Acid} \underset{\text{Hydrolysis}}{\overset{\text{Esterification}}{\rightleftharpoons}} \text{Water} + \text{Ester.}$$

The reaction to the right, esterification, is favored by the presence of concentrated sulfuric acid, which is a powerful dehydrating agent and removes the water as it is formed. When water is added to an ester, hydrolysis readily occurs, especially if either the alcohol or the acid can be removed from the system by precipitation or volatilization.

Esters tend to have pleasant scents and flavors. These properties are utilized in beverages, confections, medicines, cosmetics, and toilet waters. Some of the more commonly known esters are **amyl acetate** (banana oil) and **methyl salicylate** (oil of wintergreen). Esters are likewise utilized as very important solvents.

ETHERS

Ethers are formed by the dehydration of alcohols by concentrated sulfuric acid. Diethyl ether, $C_2H_5OC_2H_5$, which is the well-known anesthetic, is made from ethyl alcohol.

$$2\ C_2H_5OH \xrightarrow[\text{desiccant}]{(H_2SO_4)} H_2O + C_2H_5OC_2H_5.$$

The ether formed is extremely volatile and is not miscible with water. Thus it is easily distilled from the water. Ethers are extremely flammable.

PLASTICS AND SYNTHETIC FIBERS

When small organic molecules are caused to be linked together to form long chains, or complex two- and three-dimensional networks, the process is known as **polymerization,** and the products are called **high polymers.** Polymerization takes place only in molecules containing double or triple bonds, and is usually dependent upon temperature, pressure, and the presence of a suitable catalyst.

For example, the gas ethylene, $H_2C = CH_2$, when heated under pressure polymerizes to form a solid transparent plastic called polyethylene. It consists of CH_2 units linked together in an endless chain: . . . $CH_2CH_2CH_2CH_2$ . . . Ethylene is said to be the **monomer** of polyethylene.

More commonly, two or more different monomers may be mixed and then co-polymerized to form an almost endless variety of different plastic substances. Some of these plastics involve the **addition** of one monomer to another similar to the formation of polyethylene. Others involve **condensation,** or the splitting out of water molecules as the monomers link together.

In general, plastics behave toward heat in one of two ways. Either they are **thermosetting,** which means that on being heated they set to a hard infusible solid, or they are **thermoplastic,** which means that on being heated they become soft and can be remolded or reshaped. Bakelite is an example of a thermosetting plastic, while Lucite or Plexiglas is an example of a thermoplastic plastic.

Some synthetic fibers like nylon are made by co-polymerization. This substance is said to be made from coal, water, and air. Actually, hydrogen from water and nitrogen from air are united to form ammonia, which is then converted in a series of steps to a compound called **hexamethy-**

lenediamine. Phenol, extracted from coal tar, is converted by another series of steps to a compound called **adipic acid.** These two synthetic compounds are then condensed together, eliminating water and forming giant fibrous nylon molecules.

On the other hand, **rayon** is pure cellulose. In its manufacture, purified cotton fibers (virtually pure cellulose) are converted to cellulose compounds, **sodium cellulose** and **cellulose xanthate.** The latter is then added to an acid solution which regenerates pure cellulose with characteristic silky properties. Acetate rayon is the compound cellulose acetate. Thus these fibers are not related to plastics.

LIQUID CRYSTALS

Liquid crystals are jellylike organic compounds that resemble liquids in some ways and crystals in other ways, such as their ability to scatter and reflect light. Numerical displays on some watches, calculators, clocks, and automobile dashboards depend on their use.

Problem Set No. 24

1. Write the structural formula for hydroxybenzene or phenol.
2. Write the esterification-hydrolysis reaction between ethyl alcohol and acetic acid.
3. One liter of propylene weighs 1.88 g. at STP. Its composition by weight is C—85.6%, and H—14.4%. What is its molecular formula?
4. Match the compound with its class name:

1. CH_3OCH_3 a. chlorobenzene

       O
       ‖
2. $CH_3CCH_2CH_3$ b. polyethylene

3. $CH_3CH_2CH_2CH_2CH_2CH_2CH_3$ c. heptane

4. [benzene ring]—Cl   d. dimethyl ether

5. . . . $CH_2CH_2CH_2$.   e. methyl ethyl ketone

5. Given the structure of n-propyl chloride:

$$H-\underset{\underset{H}{|}}{\overset{\overset{H}{|}}{C}}-\underset{\underset{H}{|}}{\overset{\overset{H}{|}}{C}}-\underset{\underset{H}{|}}{\overset{\overset{H}{|}}{C}}-Cl$$

Write the structure for isopropyl chloride.

SUMMARY

Organic chemistry relates to the synthesis and chemistry of the **compounds of carbon,** of which over 4 million are known. Partly because the valence of carbon is four, it can make numerous compounds based on carbon-carbon and carbon-hydrogen bonds, as well as many compounds based on the carbon linkages with oxygen, nitrogen, sulfur, and other elements, and with a variety of functional groups.

Organic compounds may be **saturated** or **unsaturated.** They may also be aliphatic, aromatic, or heterocyclic.

The sources of carbon for organic chemistry are coal, petroleum, natural gas, and plants.

Organic chemical reactions proceed usually by **addition** or **substitution.**

Important classes of organic compounds include alcohols, aldehydes, ketones, acids, esters, ethers, plastics, and synthetic fibers.

CHAPTER 23

NUCLEAR CHEMISTRY

In 1896 the French scientist HENRI BEC-QUEREL accidentally discovered a strange natural phenomenon. By chance he placed a sample of a mineral called pitchblende in his desk on top of a covered photographic plate. A key happened to be between the mineral and the plate. He later used the plate, and when he developed it he found that he had a photograph of the key. The only possible explanation was that some mysterious emanation coming from the mineral was capable of penetrating the cover on the photographic plate and causing the plate to be exposed.

Shortly thereafter some European explorers in the Belgian Congo of Africa came upon a native village in which the tribal witch doctor cured his sick tribesmen by burying them to their neck in the ground around his village. It was later shown that this ground gave off the same emanation as the pitchblende.

The emanation coming from these substances was called **radioactivity.** It was quickly discovered that there were three different types of emanations coming from radioactive substances. (See Figure 53.) Pending further investigation, they were called alpha, beta, and gamma rays after the first three letters in the Greek alphabet. Further study showed that the three types of "rays" were actually the following:

1. **Alpha rays** are nuclei of helium atoms, consisting of particles containing 2 neutrons and 2 protons. Thus they are positively charged. They move at speeds ranging from 2,000 to 20,000 miles per second, or from 1 to 10 percent of the speed of light.

2. **Beta rays** are a stream of electrons. Thus they too are particles. They are, of course, negatively charged. These electrons move quite fast, almost at the speed of light.

3. **Gamma rays** are radiant energy. They are best described as bundles of energy resembling light energy, or better still, X rays. They move at the speed of light, but have neither mass nor electrical properties.

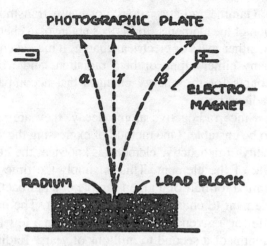

Fig. 53. Radioactivity Rays

The emanations of radioactive materials come exclusively from **nuclei of atoms.** Whether the atoms are in the free elemental state or whether they are chemically combined in compounds has no effect on their radioactivity, but the fact that the nuclei of elements undergo change during radioactivity means that a transmutation of one element into another occurs. Such a change is not an ordinary chemical change, for it involves the nucleus rather than the planetary electrons of the atom. All of the elements of higher atomic number than lead have naturally radioactive isotopes.

When the nucleus of an atom loses an alpha particle, it loses two protons, and therefore the atomic number of the remaining nucleus is two less than the atomic number of the original atoms. Thus a uranium nucleus is transformed into a thorium nucleus as the result of the loss of an alpha particle. Or a radium nucleus is transformed into a radon nucleus by the same process.

The loss of a beta particle by a nucleus of an atom involves first the decomposition of a neutron. The neutron splits into a proton and an electron. The electron leaves the nucleus, and the remaining nucleus has one more proton than the original nucleus. Thus the atomic number is increased by one unit during this type of emanation. A uranium nucleus is thus converted to a neptunium nucleus, or a neptunium nucleus is transformed into a plutonium nucleus by this process.

Gamma radiation does not cause transmutation of the elements, for it does not involve change in either mass or electrical charge. It may accompany either alpha or beta radiation, and it accounts for the energy changes that accompany radioactivity.

Since radioactive atoms decay, they are said to be unstable. One method of expressing the stability of radioactive elements is known as the **half-life** of the element. This is simply the time required for half of a given sample of a radioactive element to change to another element. The half-life for a given isotope may vary from a few billionths of a second to millions of years. Radium has a half-life of 1590 years, which means that if one has 1 gram of radium today, 0.50 grams of radium will be left 1590 years from now. After another 1590 years, 0.25 grams of the original radium will remain, and so on. It should be pointed out that these half-life periods are statistical in nature. If two atoms of radium are placed on a table, one may disintegrate immediately while the other may remain intact for millions of years. Two people may die today, or they may both live to be well over 100 years old. There is no way of knowing what will happen to a single individual. Nevertheless it is certain that of all the people born in the year 1900, practically all of them will be dead by the year 2000. But in any case, it is obvious that an element with a long half-life period is more stable than one with a short half-life.

Studies into the question of just what caused atoms to be radioactive revealed many facts concerning the nature of the atoms of the elements. Some of the highlights are summarized as follows:

1. Of the many intricate forces associated with the nucleus of an atom, two stand out. The principal nuclear particles, neutrons and protons (known collectively as **nucleons**) are held together by gravitational forces. The tremendous density of these particles (hundreds of thousands of times the density of water) and their closeness to each other is the source of this gravitational force. On the other hand, the positively charged protons exert tremendous repulsive forces on one another. The neutrons serve to buffer the effect of these repulsive forces by maintaining a space interval between the protons.

2. The attractive gravitational force tends to cause nuclear particles to combine and build up larger and larger nuclei. The repulsive electrical force tends to cause nuclei to fly apart and form smaller nuclei. It is reasonable to assume that a balance between these two forces exists somewhere between the largest and smallest nucleus. It turns out that silver (atomic number 47) has the best balance and is thus the most stable nucleus. So theoretically all elements heavier than silver should form silver by breaking up, or by **fission,** and all elements lighter than silver should form silver by combination, or by **fusion.** Elements farthest from silver on either side will liberate the greatest amount of energy as they undergo fission or fusion.

3. Another factor affecting the stability of nuclei is the ratio of the number of neutrons to the number of protons in the nucleus. For the very light atoms the ratio is 1; for atoms in the middle of the list of elements it is about 1.3; for the heavy elements it is about 1.6. If there is any marked variation from the stable ratio, the atom will be radioactive. For example: carbon 12, ^{12}C, contains 6 protons and 6 neutrons in its nucleus. Its n/p ratio is 1. It is not radioactive. But carbon 14, ^{14}C, has 8 neutrons and only 6 protons. Its n/p ratio is 1.3. This atom is radioactive because its ratio differs from the stable ratio. Thus, radioactivity is spontaneous and is independent of temperature, pressure, catalysts, or any other external conditions.

4. Both fission and fusion are different from ordinary radioactivity. Before heavy nuclei will split to form fragment nuclei of about equal size (fission), and before light nuclei will weld to form heavy nuclei (fusion), an input of energy, called **activation energy,** is required. In the case of fusion, the activation energy could be supplied by extremely high temperatures such as are found in the sun or other stars, or by the energy associated

with extremely high velocity particles at elevated temperature. In the case of fission, most nuclei also require very high activation energy. However, a few nuclei such as uranium 235 or plutonium can be activated by the energy associated with moving neutrons.

NUCLEAR REACTIONS

Investigating the nature of radioactivity, nuclear scientists discovered that it was possible to cause the artificial transmutation of elements by bombarding their nuclei with atomic particles. In general, such reactions lead to the production of artificially radioactive isotopes of the various elements. Successful transmutation has been accomplished by using the following "bullets":

1. Neutrons—^{1}n.
2. Protons (hydrogen nuclei)—^{1}H.
3. Deuterons (deuterium nuclei)—^{2}H.
4. Alpha particles (helium nuclei)—^{4}He.
5. Gamma rays (pure energy)—$^0\gamma$.

Hence we can speak of five different types of nuclear reactions. They are:

1. **Neutron-induced reactions.** These result in the formation of a heavier isotope of the bombarded element, or may result in the emission of protons or alpha particles. Examples are:

$$^{107}Ag + {}^1n = {}^{108}Ag.$$

$$^{10}B + {}^1n = {}^7Li + {}^4He.$$

$$^{14}N + {}^1n = {}^{14}C + {}^1H.$$

The last reaction is a side reaction from a-bomb blasts and can cause serious consequences. ^{14}C is intensely radioactive, has a half-life of 5,000 years, and is formed in these blasts in fairly large quantities.

2. **Proton-induced reactions.** These result in the formation of the next higher element. Neutrons may or may not be emitted during these reactions. Examples are:

$$^{12}C + {}^1H = {}^{13}N.$$

$$^{18}O + {}^1H = {}^{18}F + {}^1n.$$

The latter reaction may be used as a source of neutrons for other nuclear reactions.

3. **Deuteron-induced reactions.** These result in the formation of either protons, neutrons, or alpha particles. This type is commonly employed in the manufacture of artificially radioactive elements. Examples are:

$$^9Be + {}^2H = {}^{10}B + {}^1n.$$

$$^{23}Na + {}^2H = {}^{24}Na + {}^1H.$$

$$^{20}Ne + {}^2H = {}^{18}F + {}^4He.$$

The first reaction is commonly employed as a source of neutrons.

4. **Alpha particle-induced reactions.** These result in the emission of either protons or neutrons. Examples are:

$$^{14}N + {}^4He = {}^{17}O + {}^1H.$$

$$^{10}B + {}^4He = {}^{13}N + {}^1n.$$

$$^9Be + {}^4He = {}^{12}C + {}^1n.$$

The first reaction was the first artificial transmutation ever carried out experimentally. The last one produces nonradioactive carbon and was the original source of neutrons for experimental purposes.

5. **Gamma ray-induced reactions.** Again, these may produce either protons or neutrons. Examples are:

$$^{26}Mg + \text{gamma rays} = {}^{25}Na + {}^1H.$$

$$^9Be + \text{gamma rays} = {}^8Be + {}^1n.$$

All neutrons from the last reaction have the same energy.

The particles used for bombardment in nuclear reactions are usually energized to a definite energy level before they are permitted to strike the target nuclei. This energizing is accomplished in

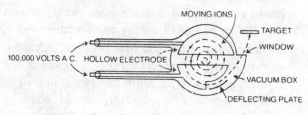

Fig. 54. The Cyclotron

various types of accelerating devices known as **cyclotrons** (Figure 54), **betatrons, synchrotrons, linear accelerators,** and various combinations of these devices. In general these devices utilize the energy of magnetic and electrical fields to speed up the bombarding particle until it has the right energy to carry out a desired nuclear reaction.

Although nuclear reactions have been described as **"atom-smashing,"** you will notice that in all of the reactions described thus far, the alpha particle, ^{4}He, is the largest subatomic particle produced. The new atoms formed are consequently only slightly smaller than the original target atoms. Therefore the reactions we have seen may more properly be described as **"atom-chipping."** Just as in ordinary chemical reactions, energy is either absorbed or released during nuclear reactions. The amount of energy associated with nuclear reactions is, in general, higher than that associated with chemical change, although some of the more powerful chemical reactions liberate a quantity of energy approaching that of nuclear reactions. However, in the late 1930s two German scientists, HAHN and STRASSMANN, discovered that when neutrons were permitted to bombard uranium, a reaction liberating unexpectedly huge quantities of energy took place, and that the products of the reaction consisted of two atoms each about half the atomic weight of uranium. This reaction was indeed "atom-smashing," for it resulted in the splitting, or **fission,** of uranium atoms into two approximately equal parts.

This fission of uranium was interesting not only because of the tremendous energy output it produced, but also because during the reaction additional neutrons were liberated which were capable of spreading the fission reaction to neighboring atoms. This fact made possible a **chain reaction** that could liberate previously unheard-of quantities of energy from relatively small amounts of matter. The first practical release of atomic energy was accomplished in the fission chain reactions of uranium and plutonium in the **atomic bombs** which brought World War II to an end. These same reactions are being carried out under carefully controlled conditions in nu-

clear reactors (see Figure 55). The heat generated is being utilized to supply steam for the generation of electrical energy at nuclear power plants. Nuclear reactors likewise are a source of radioactive materials used in medical and scientific research.

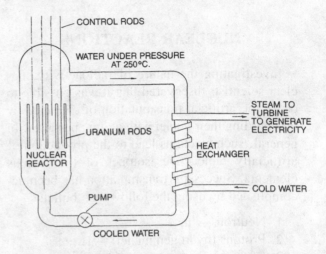

Fig. 55.

Nuclear Reactor to Make Steam to Make Electricity

About 12% of the electricity produced in the United States comes from nuclear energy. In a typical fission reaction in a nuclear reactor, ^{235}U is bombarded with neutrons and it splits or fissions into barium, krypton, and more neutrons, with an enormous release of energy. The energy is transferred by heat exchange to cold water, which is converted into steam to run a turbine that drives a generator to produce electricity. Such turbines also power nuclear submarines.

One of the interesting discoveries made in studying nuclear reactions was the finding of a new subatomic particle that is formed in some nuclear reactions. This particle, called the **positron,** has a mass equal to that of an electron, and an electrical charge equal in magnitude but opposite in sign to that of the electron. It may be thought of as a positive electron. It is usually designated by the symbol e$^+$. This particle is interesting because it can combine with an electron to form **cosmic rays.** In this reaction, the mass of both particles is transformed completely to pure energy. Positrons may also combine with neutrons to form protons.

FUSION REACTIONS

The energy we receive from the sun is believed to be caused by nuclear reactions taking place in the sun. However, these reactions are of the fusion type, involving the building up of light elements into heavier ones, rather than of the fission type involving the splitting up of heavy elements into lighter ones. In 1938 HANS BETHE of Cornell University proposed a series of nuclear reactions involving the fusion of hydrogen into helium under the catalytic influence of carbon and nitrogen. His series of reactions are:

(1) $^{12}C + {}^1H = {}^{13}N + Gamma$ rays.
(2) $\quad {}^{13}N = {}^{13}C + e^+$.
(3) $^{13}C + {}^1H = {}^{14}N + Gamma$ rays.
(4) $^{14}N + {}^1H = {}^{15}O + Gamma$ rays.
(5) $\quad {}^{15}O = {}^{15}N + e^+$.
(6) $^{15}N + {}^1H = {}^{12}C + {}^4He$.

Note that the six reactions form a complete cycle. The only net change is the fusion of 4 hydrogen atoms into a helium atom. The carbon and nitrogen are regenerated. The regeneration of nitrogen is best seen if the cycle is started at Equation 4. The positrons form cosmic radiation, and this, together with the gamma radiation, is the source of the solar energy we receive. It should be noted that the theoretical energy from this cycle is exactly equal to the measured energy radiating from the sun.

Reactions of the fusion type are also known as **thermonuclear reactions.** They require exceedingly high temperatures in order to take place. The energy liberated by this type of reaction is far greater than that associated with fission reactions. The thermonuclear reaction is the basis of the **hydrogen bomb,** which nuclear scientists have successfully discharged. Because fission reactions can be controlled, they offer a multitude of foreseeable benefits to our peaceful existence. The thermonuclear reaction, on the other hand, seems almost beyond control. It is hoped that the overwhelming destructive capacity of fusion reactions will serve as a deterrent to future wars.

In an experimental fusion reactor such as the Tokamak reactor at Princeton University, the effort is being made to fuse small nuclei on a sustainable, controlled basis to create energy for electricity and other purposes. Optimistic estimates indicate that this may be accomplished by the first two decades of the twenty-first century.

In the fusion reaction

$$^2H + {}^2H = {}^4He$$

two deuterium nuclei form an alpha particle, ^4He. About 1×10^8 kilocalories of energy are released per gram of deuterium. One kilocalorie equals 1000 calories. This is five times the energy released in fission and 10^7 times the energy released in the combustion of 1 gram of petroleum. The ocean contains 2×10^{23} grams of deuterium, enough for fusion for hundreds of years.

In Chapter 5 we were introduced to the Law of Conservation of Matter. In an ordinary chemical reaction the loss or gain in mass is too minute to be detected. Fission and fusion illustrate vividly, however, that it is mass-energy that is neither created nor destroyed in any chemical transmutation. In fission and fusion the loss in mass is sufficient to create an enormous release of energy. The relation between mass and energy is

$$E = mc^2$$

where E = energy in ergs (4.184×10^7 ergs = 1 calorie)
$\quad m$ = mass in grams
$\quad c = 3 \times 10^{10}$ cm./sec., the speed of light.

If four moles of hydrogen fuse to make one mole of helium, then $4 \times 1.008 = 4.032$ grams hydrogen make 4.00 grams helium (refer to the periodic table, Figure 12 on page 21). Note that this equation does not balance. We have lost 0.032 grams somewhere. The explanation is that this mass has been transformed into energy. Let us calculate its value:

$E = \dfrac{0.032}{4.032}$ grams lost per gram of hydrogen
$\quad \times (3 \times 10^{10})^2$
$= 0.0714 \times 10^{20}$ ergs
$= 1.71 \times 10^8$ kilocalories per gram of hydrogen reacted

which is about 10,000,000 times the energy released in burning one gram of carbon, methane, or hydrogen (see Table XLIX on page 176).

Fusion reactions are used to create transuranium elements. One atom of element 109 was discovered at the Institute for Heavy Ion Research in Darmstadt, Germany, in 1982 by fusion of an iron nucleus, by bombardment with a bismuth nucleus, with ejection of a neutron. The life of the newly created atom of element 109 was 5/1000th of a second. Element 107 was also created at the Institute for Heavy Ion Research in 1981. It is the current belief that only about twenty more elements exist and will be formed, probably by fusion. They are the "superheavy" elements of atomic number in the region of 114.

One more important fact should be pointed out. Radioactive atoms behave chemically in exactly the same manner as their nonradioactive natural isotopes. This fact is of tremendous value in research. The path of radioactive isotopes can be **traced** through complex processes which occur in the human body, in the growth of plant and animal life, and in industrial processes, with the result that much insight can be gained as to the precise nature of many of these processes. It is quite probable that through the use of tracer elements we will soon be able to solve many of the present mysteries of nature.

Problem Set No. 25

1. Complete the following nuclear reaction:

$$^{14}N + ^4He = ^{17}O + X$$

2. How does a fusion reaction differ from a fission reaction and an ordinary chemical reaction?

3. Do the emanations of radioactivity, α, β, and γ rays come from (a) the nucleus, (b) the electrons surrounding the nucleus, or (c) both nucleus and surrounding electrons?
4. Do neutron-induced reactions create heavier or lighter elements?
5. Is the energy from the sun produced by fusion or fission reactions?

SUMMARY

Nuclear chemistry involves **fission** or disintegration of the nuclei of the atoms of elements naturally or by artificial means. Natural fission is called **radioactivity.** All the actinides are radioactive. Only two of them, thorium and uranium, exist in nature in appreciable quantities.

Radioactive elements give off α, β, and γ rays. Their stability is measured by their **half-lives.**

Artificial transmutation of the elements can be effected by bombarding their nuclei with neutrons, protons, deuterons, alpha particles, and gamma rays.

Useful applications of nuclear fission include **nuclear power plants** to generate electricity. Fission releases large amounts of energy.

Nuclear chemistry also involves the **fusion** of the nuclei of light atoms such as hydrogen, with the resultant release of huge amounts of energy, such as in the hydrogen bomb. It is hoped that someday this fusion can be controlled to produce electricity for peacetime use.

CHAPTER 24

THE CHEMISTRY OF ENERGY

Chemistry is committed to developing more abundant **energy resources** at a time when world population is rapidly increasing, people are demanding higher living standards, and the most depended-on source of energy, petroleum or oil, is depleting rapidly.

SOURCES OF ENERGY

Energy can be created by physical or chemical means. **Hydroelectric energy** stored by dams on rivers and lakes, is physical. The rotation of the vanes of a windmill to create **wind energy** is also physical, as is **geothermal energy,** which taps underground beds of steam to produce electrical energy.

Most forms of energy, however, are chemical: they involve a chemical reaction, called an oxidation, to release heat. Important sources of **chemical energy** are petroleum, coal, natural gas, and, to a lesser extent, wood.

Petroleum largely contains alkanes, which can react with oxygen as follows:

$$2 C_nH_{2n+2} + (3n+1)O_2 = 2 \, nCO_2 + 2(n+1)H_2O + \text{heat energy}$$

The carbon in coal can be burned directly:

$$C + O_2 = CO_2 + \text{heat energy}$$

or it can be reacted, in the water gas reaction, with steam:

$$C + H_2O = CO + H_2 + \text{heat energy}$$

The mixture of carbon monoxide, CO, and hydrogen, H_2, is called water gas. The product, CO and H_2, can also be used as a fuel, or as a starting point, by adjusting the CO/H_2 ratio to form synthesis gas to make liquid fuels comparable to gasoline:

$$6 CO + 13 H_2 = C_6H_{14} + 6 H_2O$$

where C_6H_{14} is a form of gasoline.

Natural gas is 85% or more methane, CH_4, with some ethane, C_2H_6, propane, C_3H_8, and trace amounts of higher alkanes and carbon dioxide. The methane reacts with oxygen to release heat according to

$$CH_4 + 2 O_2 = CO_2 + 2 H_2O + \text{heat energy}$$

Wood is a complex mixture of 70–90% **cellulose,** 10–30% **lignin,** and a small percent of **wood ash.** Wood ash contains inorganic oxides left after combustion. Lignin is a polymer of repeating units of composition $C_9H_9O_5 \cdot OCH_3$. Cellulose is a polymer of repeating unit $C_6H_{10}O_5$. Wood contains some oxygen, so pound for pound its heating value is less than that of petroleum, coal, and natural gas.

Municipal garbage and plant stalks may also be combusted to yield useful heat energy.

Methanol, CH_3OH, and **ethanol,** CH_3CH_2OH, are other sources of chemical energy being studied and readied for future use. They burn as follows:

$$2 CH_3OH + 3 O_2 = 2 CO_2 + 4 H_2O + \text{heat energy}$$

$$C_2H_5OH + 3 O_2 = 2 CO_2 + 3 H_2O + \text{heat energy}$$

Methanol can be made by the destructive distillation or heating of wood in the absence of air, or it can be made from a mixture of CO and H_2 from coal or natural gas:

$$CO + 2 H_2 = CH_3OH$$

Ethanol can be made by the fermentation of carbohydrates from grain, corn, sugarcane, cassava roots, and other high starch producing plants. In the last step of fermentation a sugar, glucose, $C_6H_{12}O_6$, is formed and converted to ethanol:

$$C_6H_{12}O_6 = 2 C_2H_5OH + 2 CO_2$$

Photovoltaic cells are being developed to convert solar energy to electrical energy. The most popular photovoltaic material is ultrapure silicon. GaAs, CdS, and CdSe can also be used. The sun's rays strike an atom of silicon and knock an electron from the outer shell, starting an electrical current. Solar cells are used to power buoys, irrigation systems, communications equipment in remote areas, calculators, watches, and, mounted on rooftops, electricity for the home. It is proposed to use solar energy cells to provide the electricity for the electrolysis of water to hydrogen H_2 and oxygen O_2. The H_2 would be used as a fuel:

$$2 H_2 + O_2 = 2 H_2O + \text{heat energy}$$

Nuclear fission and **nuclear fusion** were discussed in Chapter 23. Nuclear fission involves splitting apart the nucleus of heavy atoms by bombardment with neutrons; nuclear fusion involves fusing together the nuclei of light atoms. Both processes occur with an enormous release of energy. These are not chemical reactions in the ordinary sense of involving outer shell electrons, but they are chemical reactions involving nuclear transmutations.

RENEWABLE AND NONRENEWABLE ENERGY

An energy source is either renewable or nonrenewable. It is renewable if it can be replenished. Petroleum, natural gas, and uranium for nuclear fission are nonrenewable—sooner or later their supply will become exhausted. Wood, municipal waste, plant stalks, methanol from wood, ethanol from plant sources, are renewable sources of energy. New trees and plants can be grown and consumed year after year, forever. Solar energy is also renewable. Photovoltaic cells can electrolyze water forever because the product of combustion is water itself! Silicon is made by reduction of silica—sand—an inexpensive source in superabundant quantities. Nuclear fusion is virtually unlimited energy because there is enough heavy hydrogen or deuterium in the world's oceans for fusion into helium for hundreds of years. Wind and hydroelectric energy are renewable; geothermal energy may also be.

A measure of energy consumption is the quad. One quad equals 10^{15} or one quadrillion BTU (one BTU = one British thermal unit). The United

States consumes about 79 quads per year. This rate of consumption is based on 1980 figures and is expected to increase gradually. The sources of

Table XLVIII
U.S. Energy Consumption in Quads

Energy Source	1980	Projected for 2020
Oil	34	20
Natural gas	21	15
Coal	16	71
Nuclear fission	3	18
Other (hydropower, solar, geothermal, wood, biomass and windpower)	5	19*
Total	79	143

*May also include fusion energy.

Table XLIX
Heat Liberated by Various Fuels
(By combustion with oxygen with the exception of fission and fusion)

Fuel	Energy Released, kilocalorie (kcal.)/ gram of fuel (Note: 1 BTU = 0.252 kcal.)
Natural gas	11.6
Petroleum or oil	11.3
Coal (anthracite)	7.3
Coal (bituminous)	7.0
Ethanol (C_2H_5OH)	7.1
Methanol (CH_3OH)	5.4
Wood	4.5
Hydrogen (H_2)	34.2
Carbon monoxide (CO)	2.4
Carbon (C)	7.8
Methane (CH_4)	13.3
Deuterium fusion ($2\ ^2H = {}^4He$)	1×10^8
^{235}U (uranium) fission	2×10^7

energy that supply the 79 quads are given in Table XLVIII. The heat energy released in typical combustion reactions is listed in Table XLIX.

Problem Set No. 26

1. Which of the following energy sources are renewable?

   (a) Hydroelectric
   (b) Coal
   (c) Wood
   (d) Nuclear Fission
   (e) Natural Gas

2. The complete combustion of oil, natural gas, coal, wood, and biomass yields CO_2 and H_2O. True or false?
3. In what way can solar energy provide hydrogen (H_2) fuel?
4. What percent of energy consumption in the United States is now provided by coal and nuclear fission? What is the projected percent in 2020? Why?
5. If 1 gram hydrogen (H_2) burns with oxygen to release 34.2 kcal./g. hydrogen of energy, how many BTU/mole hydrogen of energy is released?

SUMMARY

The chemistry of energy relates to the fact that most energy is produced by the combustion of **fossil fuels** (oil, natural gas, coal). **Wood** and **biomass** are also combusted. **Nuclear fission** is chemical energy in the sense that a reaction takes place, disintegrating the nucleus of ^{235}U. **Solar energy** is chemical in the sense that electrons flow into and from the outer valence shell of silicon under the influence of sunlight. Other sources of energy are physical: hydropower, geothermal, wind.

Energy is **renewable or nonrenewable** depending on whether it can be replenished. Coal and nuclear energy will become important energy sources in the future. If nuclear fusion can be harnessed by the next century, it will become significant since there is in the oceans enough deuterium to last for centuries for fusion into helium.

Convenient units of measurement of energy are the **kilocalorie (kcal.)** and **British thermal unit (BTU).** One BTU = 252 calories.

CHAPTER 25

THE CHEMISTRY OF THE ENVIRONMENT

Our **environment** consists of the **air** or atmosphere, the **water** in lakes, rivers, underground wells and streams, and oceans, and the **solid land,** or ground. This environment is now being assaulted by the products and by-products of the chemical and nuclear power industries, many of which are toxic or injurious to our health, and to the health of plants and animals. Prior to the industrial revolution which began about 1750, the chemical industry was virtually nonexistent and the chemistry of the environment was the same natural processes that had existed for thousands of years.

Toxic chemicals produced today react in chemical ways to impair the well-being of human, animal, and plant life. Their safe or safest possible disposal involves chemical or physical principles.

AIR POLLUTION

Coal contains free or combined sulfur. The **sulfur oxide gases,** SO_2 and SO_3, are produced mainly by its combustion. SO_3 can react with water vapor in the air and clouds to form sulfuric acid:

$$SO_3 + H_2O = H_2SO_4$$

The sulfuric acid dissolves in rainfall, lowering the pH to as low as 2 or 3, attacking metal surfaces, marble surfaces and statues, and plant and animal life. Metals corrode, marble disintegrates, plants and trees lose their leaves and their roots are attacked by acid rain leaching into the soil, and fish can no longer inhabit lakes acidified by **acid rain.**

Marble disintegrates according to

$$CaCO_3 + H_2SO_4 = CaSO_4 + CO_2 + H_2O.$$

Scrubbing systems containing alkaline materials such as MgO are being placed in coal-burning electric plants to remove the sulfur oxides before they enter the air:

$$2\,MgO + 2\,SO_2 + O_2 = 2\,MgSO_4$$

Carbon monoxide is a toxic gas that comes principally from tailpipes of cars from the incomplete combustion of gasoline:

$$2\,C_8H_{18} + 17\,O_2 = 16\,CO + 18\,H_2O$$

Since 1975 American-made cars have been equipped with **catalytic converters** containing catalyst pellets coated with platinum or palladium to oxidize the CO in the tailpipe:

$$\overset{Pt,Pd}{2\,CO + O_2 = 2\,CO_2}$$

Car engines and coal power plants also oxidize nitrogen in the intake air to **nitric oxide, NO,** and **nitrogen dioxide, NO_2.** NO_2 is a yellow to reddish-brown gas that causes smog. It also reacts with water vapor in the air to form nitric acid:

$$3\,NO_2 + H_2O = 2\,HNO_3 + NO.$$

Oxygen atoms, O, and **ozone, O_3,** are also constituents of **smog.** They cause eye irritation. They are produced as follows:

$$NO_2 = NO + O$$

$$O_2 + O = O_3$$

CHEMISTRY OF TRACE METALS

Trace metals are present in chemical products and in nature. They find their way into water and the soil. Some trace metals are found in food; derived from water and plants and animals, they are essential to health. Other trace metals in the environment may be toxic; they include beryllium, cadmium, mercury, and lead. Beryllium can cause lung disease; cadmium can cause high blood pressure and affect the heart.

If **mercury** is discarded on lake or river shores, or in lakes or rivers, microorganisms can convert the mercury into toxic dimethyl mercury, $(CH_3)_2Hg$, which finds its way into fish and into people who eat the fish.

Tetraethyl lead, $(C_2H_5)_4Pb$, along with dichloroethane, $C_2H_4Cl_2$, and dibromoethane, $C_2H_4Br_2$, is added to some gasolines to prevent knocking. Newer cars are now designed to run on lead-free gasoline. The auto exhaust from leaded gasoline emits particles of $PbClBr$ and $NH_4Cl \cdot 2\,PbClBr$ which are small enough to be breathed in and retained by the lungs. The use of $(C_2H_5)_4Pb$ is decreasing each year.

Lead pigments in paints such as $Pb(OH)_2$, $PbCO_3$, and Pb_3O_4 are used less in paints than heretofore because of their toxicity.

Concentrations of trace metals in the ground, water, plants, and animals are expressed in **parts per million, ppm,** which means 1 gram metal in 10^6 grams host material. Fish normally contain 0.2 ppm or less of mercury, or

$$0.2 \times 10^{-6}\ g. = 2 \times 10^{-7}\ g.\ \text{mercury}$$
$$\text{per gram of fish.}$$

HAZARDOUS WASTES IN THE GROUND AND WATER

Over 76 billion pounds of hazardous waste are generated each year in the United States. Their source is given in Table L, based on 1978 statistics. Proper disposal usually involves incineration or systematic placement in a sanitary landfill. About 10% are properly disposed of as shown in Table LI, based on 1978 figures.

There is growing concern on the use of **landfills.** Even the most secure landfills will eventually leak, oozing their contents into underground streams or into the surrounding surface environment. Alternatives to sanitary landfills are:

1. Reduce generation of waste products by industry.
2. Recover and recycle waste materials.
3. Incinerate properly on land or at sea.
4. Reduce volume and toxicity of waste by physical, chemical, and biological treatments.
5. Inject waste into deep wells.
6. Provide interim storage in surface tanks.
7. Prohibit products or processes that create wastes that in terms of safety are unmanageable.

Table L
Hazardous Wastes Generated by Industry

Source	Billion Pounds
Organic chemicals	26
Primary metals	29
Electroplating	9
Inorganic chemicals	9
Textile, oil refinery, rubbers, and plastics	3
Total	76

Hazardous wastes include pesticide residues, halogenated hydrocarbons such as polychlorobiphenyls (PCBs), asbestos, acids and bases, and a host of other chemicals. Their disposal in any body of water is improper. Fresh water is eventually used by humans, plants, and animals. Ocean water is a reservoir for marine life, some of which becomes food for humans.

Table LI
Disposal of Hazardous Wastes

Method	Billion Pounds
Proper disposal	
On-site disposal	5
Secure landfill	2.6
Total	7.6
Improper disposal	
Unlined lagoons and ponds	34
Nonsecure landfills	20
Ocean dumping, sewers, roads, deep wells, ordinary incinerators	14.4
Total	68.4
Grand Total	76.0

Solid wastes may be classified as municipal, industrial, mineral, and agricultural. The last three are disposed of by their producers. Municipal wastes are disposed of in landfills, proper incineration by oxidation with air at 1050–1250 K, or pyrolysis in the absence of air. Solid waste may also be **reclaimed** for use in a different way from its original use; **recycled,** as scrap iron is converted to steel; or **reused,** as with beverage bottles.

Municipal and industrial waste waters must be treated physically and chemically. In municipal waste water the suspended solids are allowed to settle. The chemically organic matter is then oxidized with the aid of microorganisms to remove **biological oxygen demand, BOD.** BOD is a measure of the amount of organic material in the waste water.

RADIOACTIVE WASTES

Radioactive wastes from nuclear power plants are estimated to be about 17 million pounds, from the start of nuclear power in 1957 to 1983. They have been stored in ponds of water adjacent to nuclear power plants. A permanent storage for them is proposed by the **Nuclear Waste Policy Act of 1982** in metal canisters stored 2,000 to 4,000 feet underground in solid-rock cavern formations that are sufficiently stable to keep the radioactive waste safe and dry for at least 10,000 years.

Problem Set No. 27

1. The highest concentration of lead allowed in water is 0.05 ppm. How many grams of lead is this, per gram of water?
2. Smog comes mostly from (a) mining operations, (b) automobiles, (c) chemical plants, (d) none of these.
3. If the pH of rain in your area is less than 7, what does this signify?
4. A secure landfill has been thought to be a proper way to dispose of hazardous wastes. Why?
5. Use can be made of lime, CaO, to remove SO_2 and SO_3 from stack gases in coal-fired electric plants. True or false?

SUMMARY

The **environment** consists of the **air, water,** and **solid land** at the earth's surface. Chemicals produced by industry and their by-products may be toxic or hazardous and they must not find their way into the air we breathe, the water we drink, or the soil from which plants grow and animals find nourishment. They must not be allowed to enter the oceans.

Toxicity of chemicals is also chemical in nature, involving reactions with different areas of the body, causing injury.

More than ever we are becoming aware that toxic chemicals must be properly disposed of. Methods of disposal involve both physical and chemical means. This also applies to the radioactive products of nuclear reactions at nuclear power plants.

CHAPTER 26

BIOCHEMISTRY

Biochemistry is the **organic chemistry of life.** The Greek word *bios* means life.

The major chemical building blocks or substances of life are **carbohydrates, fats, proteins, and nucleic acids.**

CARBOHYDRATES

Carbohydrates are compounds of carbon, hydrogen, and oxygen which usually have a hydrogen to oxygen atomic ratio of 2 to 1, the same as in water. Carbohydrates are also known as **saccharides.** The chief members of this family of compounds are **cellulose, starches,** and **sugars.** Both cellulose and starch are formed in plants from carbon dioxide and water by photosynthesis. Both may be represented by the following formula: $(C_6H_{10}O_5)_x$. This formula indicates that the molecules of these compounds consist of a series of units, each containing the basic carbon-hydrogen-oxygen ratio indicated, joined to form a long chain. The actual number of such units present in a single molecule has not yet been discovered. Cellulose differs from starch in the spatial arrangement of the atoms in the molecule, and contains more of the basic units than starch. They are both referred to as **polysaccharides** because they contain more than one simple unit. Starch is obtained chiefly from cereal grains and from potatoes. Cellulose is the basic constituent of plant structures. Cotton and linen are almost pure cellulose, while wood is cellulose mixed with compounds called **lignins.**

Polysaccharides hydrolyze in the presence of acid and enzymes to form simple sugars of the formula $C_6H_{12}O_6$. There are many different sugars with this formula. They differ from one another again in the arrangement of the atoms in their molecules, and thus have different properties. The two basic types of simple sugars are **glucose** (also called **dextrose**), and **fructose** (also called **levulose**). Note the structures of these two compounds:

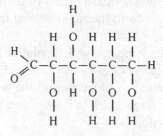

Glucose

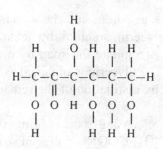

Fructose

You can see that glucose is a polyhydroxy aldehyde while fructose is a polyhydroxy ketone. Simple sugars are called **monosaccharides.** Our bodies have the enzymes necessary to hydrolyze starch to glucose, but lack the enzymes required to digest cellulose.

Ordinary cane sugar is **sucrose,** obtained from sugarcane or sugar beets. It is a **disaccharide,** consisting of two combined basic units of the formula $C_{12}H_{22}O_{11}$. In hydrolysis it adds one molecule of water to form one molecule of glucose and one molecule of fructose. This equimolar mixture is known as **invert sugar.** Honey is essentially invert sugar.

FATS

A fat is an ester of glycerin and an organic acid. Butter fat is a mixture of such compounds whose acid radicals contain 4 or more carbon atoms. The chief compound in beef fat is **glycerin tristearate,** $(C_{17}H_{35}COO)_3C_3H_5$. Vegetable oils used in cooking, such as cottonseed oil, olive oil, soy bean oil, and coconut oil, all contain unsaturated esters of glycerin which, when hydrogenated in the presence of a nickel catalyst, yield solid fats. Fats are widely used in cooking, particularly as shortening, and in the making of candles. They form the raw material from which soap is made.

When a fat is heated with a solution of sodium hydroxide, hydrolysis takes place and the fat ester is converted to glycerin and the sodium salt of the acid. Such a salt is known as a **soap.** The general reaction is:

$$\text{Fat} + \text{Alkali Solution} = \text{Glycerin} + \text{Soap}$$

Since soap is a sodium salt, the soap is separated from the glycerin solution by adding sodium chloride to precipitate the soap by the common ion effect. The glycerin is obtained as a valuable by-product by distillation of the remaining solution.

Ordinary cleansing soap is primarily **sodium stearate,** $C_{17}H_{35}COONa$. In facial soaps the excess alkali is removed, but in laundry soaps more alkali is added to provide more grease-cutting action. Floating soaps are aerated to lower the specific gravity below 1. Shaving soaps are made with potassium hydroxide rather than sodium hydroxide. Soaps may also contain perfumes and other germicidal or cleansing additives.

EXPERIMENT 30: Prepare various solutions of soap and water by shaking different kinds of soap with about 1 cup of water in each case. Be sure to include both face soap and laundry soap. Add a few drops of phenolphthalein solution (prepared in Experiment 5) to each soap solution. The intensity of the pink or red color is an indication of the amount of free alkali in each type of soap. Facial soaps should contain less free alkali than laundry soaps. Do your results concur?

In addition to soaps, there are **detergents.** Detergent comes from the Latin *detergere*—to cleanse. Like soaps, they contain a long-chain aliphatic group attached to a sodium salt of an acidic group. A widely used detergent is:

$$CH_3CH_2CH_2CH_2CH_2CH_2CH_2CH_2CH_2CHCH_2CH_3$$

with a benzene ring bearing SO_3Na

In 1935 there were no detergents. They have been developed since then from petroleum, instead of from animal fat, as a substitute for soap. They now have an 88% market share compared to 12% for soap. An advantage of detergents over soap is that hard water contains Mg^{++} and Ca^{++} ions. They form insoluble salts with soap, such as the insoluble calcium stearate, $Ca(C_{17}H_{35}O_2)_2$. The calcium and magnesium salts of detergents, on the other hand, are not insoluble.

PROTEINS

Proteins are **complex polymer molecules** that are the building blocks of animal tissues. Also, hair, skin, nails, and feathers are mostly protein. In hydrolysis they yield the **amino acids** that make up the protein.

An amino acid has the general structure:

$$R-\underset{\underset{NH_2}{|}}{CH}-\overset{\overset{O}{\|}}{C}-OH$$

where R is a hydrocarbon group that may contain oxygen or sulfur atoms. There are twenty "most common" amino acids. **Glycine** is the simplest. Its formula is:

$$H-\underset{\underset{NH_2}{|}}{CH}-\overset{\overset{O}{\|}}{C}-OH$$

Among the most common amino acids are the **essential amino acids** which are required by humans for health and growth but which the human body itself cannot make. They must be provided in the foods we eat.

The linkage in the polymers of the amino acids is the **peptide bond.** This bond is:

$$-\overset{\overset{H}{|}}{N}-\overset{\overset{O}{\|}}{C}-$$

A protein that contains many of these bonds is a polypeptide. Three typical protein units of a polypeptide are:

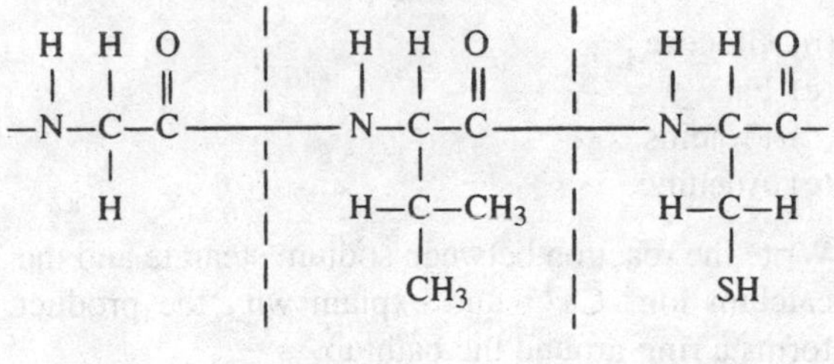

At the top left of the page are chemical structure diagrams showing amino acid chains (peptide bonds with side chains including CH₃, and SH groups).

NUCLEIC ACIDS

Nucleic acids are a group of **protein-combined polymers** found in all living cells, both plant and animal. Important nucleic acid polymers are **deoxyribonucleic acid, DNA,** and **ribonucleic acid, RNA.** DNA is found mainly in the nucleus of all cells. RNA is found mainly elsewhere in the cell. They participate in the biosynthesis of proteins. DNA carries the genetic information that allows a species to reproduce its own kind. It is DNA that determines whether a living thing is a robin or a bluefish, a tree or a human.

DNA is formed in the shape of a **double helix,** as shown in Figure 56.

The two strands or chains are composed of two **pentose sugar-phosphate chains** that form the helix. They are held together by the hydrogen bonding between the base pairs or **nucleotide pairs** adenine-thymine and guanine-cytosine. Each of the ladder rungs is a pair of complementary nucleotides. Many nucleotide pairs are grouped into triplets called **codons.** Each codon can direct the formation of one of twenty amino acids. A few hundred to several thousand codons make up a **gene.** Humans have over 100,000 genes. They are arranged into larger DNA molecules known as **chromosomes.** The constituents of DNA are shown in Figure 57.

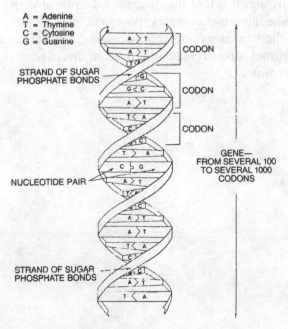

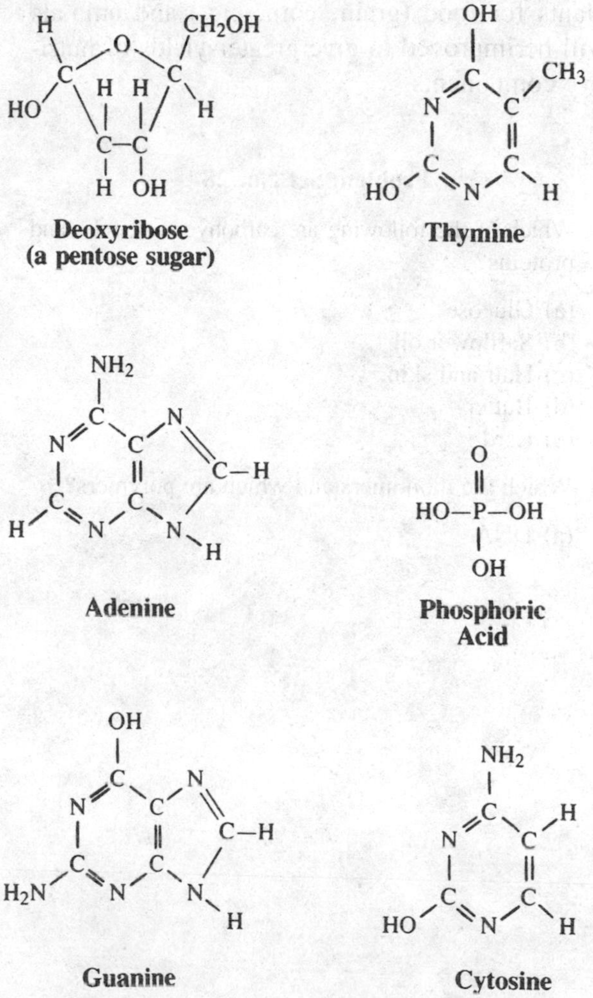

A = Adenine
T = Thymine
C = Cytosine
G = Guanine

STRAND OF SUGAR PHOSPHATE BONDS

CODON

CODON

CODON

GENE—FROM SEVERAL 100 TO SEVERAL 1000 CODONS

NUCLEOTIDE PAIR

STRAND OF SUGAR PHOSPHATE BONDS

Fig. 56. The DNA Double Helix

Deoxyribose (a pentose sugar)

Thymine

Adenine

Phosphoric Acid

Guanine

Cytosine

Fig. 57. Constituents of DNA

In 1978 synthetic genes were created in the chemical laboratory to make **human insulin** by recombinant DNA (rDNA) techniques. The genes were placed in bacteria, and the bacteria produced human insulin. Insulin is a protein of molecular weight of about 6000. Pharmaceutical factories have been constructed to make human insulin by this method, to treat people afflicted with diabetes. This represents a new, inexhaustible source, independent of the limited supply of insulin previously found in the pancreas glands of cattle and swine.

By recombinant DNA technology it is expected that more useful pharmaceutical chemical products will be made and that the varieties of plants for food (grain, corn, etc.) and animals will be improved to give greater yields of nutritive components.

Problem Set No. 28

1. Which of the following are carbohydrates, fats, and proteins?

   (a) Glucose
   (b) Safflower oil
   (c) Hair and skin
   (d) Butter
   (e) Lard

2. Which are monomers and which are polymers?

   (a) DNA
   (b) Glucose
   (c) Fat
   (d) Proteins
   (e) Adenine

3. Write the reaction between sodium stearate and the calcium ion, Ca^+, and explain why the product forms a ring around the bathtub.

4. Fats are found in

   (a) Animals
   (b) Plants
   (c) Animals and plants

5. What is the basic group of an amino acid?

SUMMARY

Carbohydrates, fats, proteins, and **nucleic acids** are the building blocks of plant and animal life. They are organic compounds and polymers synthesized by plants and animals, ranging from single-cellular life, such as bacteria, to multi-celled organisms such as the giant redwood trees of California.

Recombinant DNA (rDNA) technology in the chemical laboratory can produce genes synthetically and is a tool that is expected to be able to make important complex bioorganic compounds in the laboratory and pharmaceutical production facilities, and to create improved strains of plants and animals.

CHAPTER 27

CHEMISTRY OF THE MOON AND THE PLANETS

The first man-made earth satellite or celestial body, *Sputnik 1,* was placed in earth orbit by the U.S.S.R. on October 4, 1957. On January 31, 1958, *Explorer 1,* the first U.S. satellite, was launched. These events were the dawn of the **space age.**

Since then an incredible program has placed men on the moon and numerous satellites in orbit around the earth, and has launched spacecrafts to visit Jupiter, Saturn, and the other outer planets (see Figure 58). Spacecraft have landed on Venus and Mars. The chemical composition of the planets' and moon's atmosphere and surface have been studied. Already plans are being conjectured to mine the minerals on the moon for use on earth and for the construction of colonies in space. Suddenly the realm of chemistry has been extended millions of miles from earth. Earth may someday no longer be man's only habitat. In addition to the moon, the inner planets may be minable, since the inner planets—Mercury, Venus, Earth, and Mars—are made of **rocky materials.** The huge outer planets—Jupiter, Saturn, Uranus, and Neptune—are made of **liquefied light gases.** The composition of Pluto remains unknown.

In addition to the planets, there are hundreds of **asteroids** forming the asteroid belt between Mars and Jupiter. They are small, rocky planets ranging in size from large rocks to the small planet Ceres, 635 miles in diameter. All of these may be minable also.

The surface temperature, pressure, and atmospheric composition of the planets and the moon are given in Table LII.

CLIMATE OF THE PLANETS

The climatic conditions for life as we know it on earth are nonexistent on the moon and planets. The temperatures and pressures are harsh—either too high or too low. The atmospheres contain no

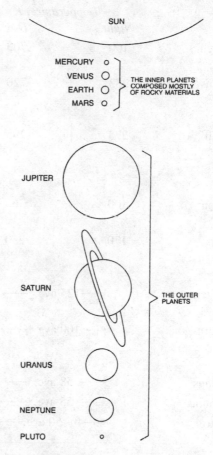

Fig. 58. The Planets

oxygen and are poisonous or inert. There is no water to drink or to support life. See Table LII.

THE MOON

The lunar surface is composed of the oxides principally of silicon, aluminum, titanium, chromium, iron, manganese, magnesium, and calcium, with trace amounts of the oxides of sodium and potassium. The composition varies from the Highlands regions to the Lowlands or Mare regions. The oxides are listed in Table LIII.

Table LII
The Atmosphere of the Planets and the Moon

Planet	Average Surface Temperature, K		Surface Pressure, atm.	Atmospheric Composition
	Night	Day		
Mercury	100	700	10^{-14}	H_2, He, and other solar wind gases
Venus	233	720	93	96% CO_2, 3.5% N_2, trace noble gases, H_2O, O_2, CO, SO_2
Earth	275	295	1	78% N_2, 21% O_2, 1% Ar, trace CO_2, H_2O, and noble gases in addition to Ar
Moon	100	365	10^{-14}	H_2, He, and other solar wind gases
Mars	170	250	6×10^{-3}	95.3% CO_2, 2.7% N_2, 1.6% Ar, trace H_2O, O_2, CO, noble gases
Jupiter	←——125 ave.*——→		10^{4}**	90% H_2, 10% He, trace CH_4, H_2O, NH_3
Saturn	←——100 ave.*——→		10^{3}**	>94% H_2, <6% He, trace CH_4, NH_3, H_2, C_2H_6, H_2O
Uranus	←—— 58 ave.*——→		2–3**	H_2, He, CH_4
Neptune	←—— 55 ave.*——→		5–10**	H_2, He, CH_4
Pluto	←——40–50 ave.——→		10^{-3}	CH_4 and perhaps Ne

*Average for cloud tops or upper atmosphere; surface not clearly defined.

**Very hypothetical; surface not clearly defined.

Table LIII
The Oxides of the Moon

SiO_2	Cr_2O_3	MgO	Na_2O
Al_2O_3	FeO	CaO	K_2O
TiO_2	MnO		

It is believed by scientists that these ores can be mined and processed to their metals on the moon or at space stations in deep space. In addition to metals, glass would be produced, and oxygen for breathing. One method to win the metals from their ores might be **electrolysis,** using solar heat energy—the oxygen and metals separating at anode and cathode. The freed oxygen could be used for breathing or converted to water, with hydrogen brought from the earth.

The metallic construction materials could then be fashioned into huge **space stations** or colonies at about 250,000 miles from earth, each containing a population of about 10,000 people.

ANSWERS TO PROBLEM SETS

Problem Set No. 1

1. From Table I, 1 fl. oz. = 29.573 ml. Therefore 3 fl. oz. = 88.719 ml.

2. From Table I, 1 lb. = 453.6 g.; 1 oz. = 28.35 g. Finding sum of 1 lb. as grams and 3 oz. as grams, and moving decimal three places to left we have: 1 lb. 3 oz. = 0.53865 Kg.

3. From Table I, 1 sq. in. = 6.4516 sq. cm. Multiplying by 528 and moving decimal 4 places to left (2 unit changes) we have: 528 sq. in. = 0.3406 sq. m.

4. 22,400 cc. = 0.022400 c.m. From Table I, 1 c.m. = 35.3 cu. ft. Therefore 0.022400 c.m. = 0.7907 cu. ft.

5. From Table I, 1 g. = 0.0353 oz. Therefore 250 g. = 8.825 oz.

6. From Table I, 1 sq. yd. = 0.8361 sq. m. Multiplying by 25 and moving decimal 4 places to right (2 unit changes) we have: 25 sq. yd. = 209,025 sq. cm.

7. From Table I, 1 fl. oz. = 29.573 ml. Multiplying by 8 and moving decimal 3 places to left we have: 1 cup = 8 fl. oz. = 0.237 l.

8. From Table I, 1 fl. oz. = 29.573 ml. Therefore 1 tablespoon = 0.5 fl. oz. = 14.787 ml.

9. From Table I, 1 gal. = 3.7853 l. Therefore 18 gal. = 68.135 l.

10. Multiplying the dimensions, we have 58.44 cu. in. in the book. From Table I, 1 cu. in. = 16.387 cc. Multiplying by 58.44 and moving decimal 3 places to left (1 unit change) we have: volume of book = 0.958 c. dm.

Problem Set No. 2

1. (a) Specific; (b) Accidental; (c) Specific

2. From Table III, density of aluminum is 2.7 g./cc.

   Volume of 5.4 g. of aluminum = $\dfrac{5.4 \text{ g.}}{2.7 \text{ g./cc.}}$ = 2 cc.

3. From Table III, 1 cc. of gold weighs 19.3 g. and density of cork is 0.22 g./cc. Therefore 19.3 g. of cork would occupy $\dfrac{19.3 \text{ g.}}{0.22 \text{ g./cc.}}$ = 87.7 cc.

4. Specific gravity = density in g./cc. Density in g./cc. × 62.4 = Density in lbs./cu. ft. Therefore: 1 cu. ft. weighs: 1.28 × 62.4 = 79.9 lbs.

5. (a) Those with less density: cork, ice.
   (b) Those with greater density: gold.

6. Carbon, hydrogen, oxygen, phosphorus, potassium, iodine, nitrogen, sulfur, calcium, iron, sodium, chlorine.

7. (a) mixture; (b) substance; (c) mixture; (d) mixture; (e) substance.

8. (a) physical; (b) chemical; (c) chemical; (d) physical; (e) physical.

9. (a) combination; (b) decomposition; (c) replacement; (d) double displacement.

10. Skin. The bubbles first form at the point of contact between the hydrogen peroxide solution and the skin.

Problem Set No. 3

1. The k-shell is number 1, the l-shell is number 2, etc. The relationship to use is:

   $$\# = 2\,s^2.$$

   Therefore:

   (a) $2 \times 1^2 = 2$ *electrons* in k-shell.
   (b) $2 \times 2^2 = 8$ *electrons* in l-shell.
   (c) $2 \times 3^2 = 18$ *electrons* in m-shell.
   (d) $2 \times 4^2 = 32$ *electrons* in n-shell.
   (e) $2 \times 5^2 = 50$ *electrons* in o-shell.

2. (a) Nucleus: 5 p, 6 n; Electrons: 2, 3.
   (b) Nucleus: 12 p, 12 n; Electrons: 2, 8, 2.
   (c) Nucleus: 18 p, 22 n; Electrons: 2, 8, 8.
   (d) Nucleus: 34 p, 45 n; Electrons: 2, 8, 18, 6.
   (e) Nucleus: 40 p, 53 n; Electrons: 2, 8, 18, 10, 2.

Problem Set No. 4

1. Rule 1, p. 26, tells us that the valence number of each sodium atom is +1. Each oxygen atom has received one electron from a sodium atom and shares one electron with another oxygen atom. Rule 5a tells us that the oxygen-to-oxygen bond doesn't count, so the valence number of each oxygen atom

must be -1. Rule 5b tells us that oxygen has a valence number of -1 only in peroxides, so the name of this compound is *sodium peroxide*.

2. Al, a Group III element, has a valence number of $+3$. S, a Group VI element, has a valence number of -2. Crisscrossing and writing as subscripts, we have: Al_2S_3. Its name is *aluminum sulfide*.

3. Mg, a Group II element, has a valence number of $+2$. N, a Group V element, has a valence number of -3. Crisscrossing and writing as subscripts, we have: Mg_3N_2. Its name is *magnesium nitride*.

4. Each Na atom is $+1$. Total: $+2$. Each O atom is -2. Total: -6. Therefore, the C atom must be: $+4$.

5. The Ca atom is $+2$. Total: $+2$. The NO_3 ion is -1. Total: -2 (See Table XII). The problem here, then, is to show how the nitrate radical has a net valence number of -1. Each O atom is -2. Total: -6. Therefore the N atom must be $+5$ if the radical is to have 1 excess negative charge.

6. The NH_4 ion is $+1$. The SO_4 ion is -2. Crisscrossing and writing as subscripts, we have: $(NH_4)_2SO_4$.

7. Mg has a valence number of $+2$. The PO_4 ion is -3. Crisscrossing and writing as subscripts, we have: $Mg_3(PO_4)_2$.

8. Atomic wt. of C is 12.0 $(\times 12) = 144.0$
   Atomic wt. of H is 1.0 $(\times 22) = 22.0$
   Atomic wt. of O is 16.0 $(\times 11) = \underline{176.0}$
   Formula wt. of sugar   342.0

9. Atomic wt. of Al is 27.0 $(\times 2) = 54.0$
   Atomic wt. of S is 32.1 $(\times 3) = 96.3$
   Atomic wt. of O is 16.0 $(\times 12) = \underline{192.0}$
   Formula wt. of aluminum sulfate   342.3

10. Formula wt. of NaOH is $23.0 + 16.0 + 1.0 = 40.0$. Using Equation (1) on p. 29, we have:

$$\frac{200}{40.0} = 5.0 \text{ gram-moles of NaOH.}$$

Problem Set No. 5

1. By Avogadro's Hypothesis, the equal volumes contain the same number of molecules. Therefore the nitrogen molecule weighs 14 times as much as the hydrogen molecule. Since hydrogen, H_2, has a molecular weight of 2, the molecular weight of nitrogen must be:

$$2 \times 14 = 28.$$

Since the atomic weight of nitrogen is 14, the number of atoms in the nitrogen molecule must be:

$$\frac{28}{14} = 2.$$

Therefore the formula of nitrogen gas is: N_2.

2.

	No. of Atoms	Atomic Weight	Total Weight
Carbon:	12	12	144
Hydrogen:	22	1	22
Oxygen:	11	16	<u>176</u>
Molecular weight of sugar:			342

$$\% \, C = \frac{144}{342} \times 100 = 42.10\%$$

$$\% \, H = \frac{22}{342} \times 100 = 6.44\%$$

$$\% \, O = \frac{176}{342} \times 100 = 51.46\%$$

3. $\% \, Cu = \dfrac{63.5}{249.6} \times 100 = 25.44\%$

4. First, find % Cl in pure NaCl. $\% \, Cl = \dfrac{35.5}{58.5} \times 100$
   $= 60.68\%$

   $\% \text{ purity} = \dfrac{58\%}{60.68\%} \times 100 = 95.58\%$ pure NaCl.

5. Aluminum: $\dfrac{52.9}{27.0} = 1.96; \dfrac{1.96}{1.96} = 1$

   Oxygen: $\dfrac{47.1}{16.0} = 2.94; \dfrac{2.94}{1.96} = 1.5.$

To produce whole numbers, we multiply each result by 2. The formula of the compound is therefore: Al_2O_3.

6. Copper: $\dfrac{50.88}{63.6} = 0.8; \dfrac{0.8}{0.4} = 2.$

   Sulfur: $\dfrac{12.84}{32.1} = 0.4; \dfrac{0.4}{0.4} = 1.$

Formula is: Cu_2S.

7. (a) $2\,NaCl + H_2SO_4 = Na_2SO_4 + 2\,HCl.$
   (b) $4\,NH_3 + 3\,O_2 = 2\,N_2 + 6\,H_2O.$
   (c) $2\,ZnS + 3\,O_2 = 2\,ZnO + 2\,SO_2.$
   (d) $C_3H_8 + 5\,O_2 = 3\,CO_2 + 4\,H_2O.$
   (e) $Ca_3(PO_4)_2 + 3\,SiO_2 + 5\,C = 3\,CaSiO_3 + 5\,CO + 2\,P.$

8. $CaCO_3 = CaO + CO_2.$

$$\begin{array}{cc} 100.1 & 56.1 \\ \dfrac{500}{100.1} = & \dfrac{x}{56.1} \end{array}$$

$$x = \frac{500 \times 56.1}{100.1}$$
$$x = 280 \text{ lbs. of lime.}$$

9. (a) $2 KNO_3 = 2 KNO_2 + O_2$.

(b) 2 moles of KNO_3 produce 1 mole of O_2. Therefore 12 moles of KNO_3 will produce 6 moles of O_2.

(c) Potassium Nitrite.

(d) 12 moles of KNO_3 produce 12 moles of KNO_2. One gram-mole of KNO_2 is: $39.1 + 14.0 + 32.0 = 85.1$ g. of KNO_2. Therefore $12 \times 85.1 = 1021.2$ g. of KNO_2 produced.

10. (a) Moles of N_2 admitted: $\frac{280}{28} = 10$.

Moles of H_2 admitted: $\frac{100}{2} = 50$.

Ratio of moles of H_2 to moles of $N_2 = 50:10 = 5:1$.

Molar ratio required by the equation $= H_2:N_2 = 3:1$.

Therefore there is an excess of hydrogen, and nitrogen is the limiting reactant.

(b) Moles of N_2 admitted $= 10$ moles (found above).

(c) Moles of H_2 required $= 3 \times 10 = 30$ moles. Moles of excess $H_2 = 50 - 30 = 20$ moles.

(d) Since each mole of N_2 produces 2 moles of NH_3, the moles of NH_3 which can be produced $= 10 \times 2 = 20$ moles of NH_3.

(e) Molecular weight of $NH_3 = 14 + 3 = 17$. Weight of NH_3 which can be produced $= 20 \times 17 = 340$ grams.

Problem Set No. 6

1. $F = 9/5 \, C + 32$; $F = \frac{9 \times 30}{5} + 32$; $F = 86° F$.

2. $F = 9/5 \, C + 32$; $68 = 9/5 \, C + 32$; $C = 5/9 \, (68 - 32)$; $C = 20° C$.

3. $C = 5/9 \, (F - 32)$; $C = 5/9 \, (95 - 32)$; $C = 35°$; $K = 35 + 273 = 308$ K.

4. $V_2 = 500 \times \frac{770}{1540} = 250$ cc.

5. $V_2 = 350 \times \frac{461}{320} = 504$ cc.

6. $P_1/T_1 = P_2/T_2$ (by canceling constant V from Combined Gas Law)

$$P_2 = 750 \times \frac{525}{293} = 1344 \text{ torr}$$

7. $750 \times 0.108 = 81$ torr of CO_2; $750 \times 0.022 = 16.5$ torr of CO; $750 \times 0.045 = 33.8$ torr of O_2; $750 \times 0.825 = 618.7$ torr of N_2.

8. $V_2 = 150 \times \frac{273}{295} \times \frac{740}{760} = 135$ cc.

9. $V_2 = 330 \times \frac{303}{273} \times \frac{29.92}{30.40} = 360$ cc.

10. $V_2 = 400 \times \frac{273}{297} \times \frac{(767.4 - 22.4)}{760} = 360$ cc.

Problem Set No. 7

1. Volume of gas at standard conditions:

$$V_2 = 555 \times \frac{273}{295} \times \frac{740}{760} = 500 \text{ ml.}$$

$$\text{Density} = \frac{0.6465}{500} = 0.001293 \text{ g./ml.}$$

2. From Table XV: Density of He $= 0.00017847$
Density of $Cl_2 = 0.003214$
By Law of Diffusion:

$$\frac{\text{Chlorine rate}}{\text{Helium rate}} = \sqrt{\frac{0.00017847}{0.003214}} = \sqrt{\frac{1}{18}} = \frac{1}{4.2}$$

Helium diffuses 4.2 times faster than chlorine.

3. Molecular weight of hydrogen is 2.016. Weight of a hydrogen molecule is:

$$\frac{2.016}{6.022 \times 10^{23}} = 3.34 \times 10^{-24} \text{ g.}$$

A hydrogen atom thus weighs half this, or 1.67×10^{-24} g.

4. Volume of gas at standard conditions:

$$V_2 = 82.2 \times \frac{273}{390} \times \frac{740}{760} = 56.0 \text{ ml.}$$

Weight of 22.4 liters would be:

$$\frac{0.24}{56} = \frac{x}{22400}$$

$x = 96$ g. which is the molecular weight.

5. Volume of gas at standard conditions:

$$V_2 = 523 \times \frac{273}{302} \times \frac{733}{760} = 456 \text{ ml.}$$

Molecular weight would be:

$$\frac{0.855}{456} = \frac{x}{22400}$$
$$x = 42$$

Simplest formula of compound would be:

Carbon: $\frac{85.72}{12} = 7.143$; $\frac{7.143}{7.143} = 1$.

Hydrogen: $\frac{14.28}{1} = 14.28$; $\frac{14.28}{7.143} = 2$.

Simplest formula is therefore CH_2. Formula weight of simplest formula is $12 + 2 = 14$. Multiplier of subscripts must be: $42/14 = 3$. True formula must then be: C_3H_6.

6. The equation for the reaction is:

$$Zn + 2\ HCl = ZnCl_2 + H_2.$$

Thus each mole of zinc produces 1 mole of H_2. Moles of zinc $= 10/65.4 = 0.153$ moles. Moles of H_2 are: 0.153. Volume of H_2 at standard conditions:

$$0.153 \times 22.4 = 3.43 \text{ liters.}$$

7. The equation for the reaction is:

$$CaCO_3 = CO_2 + CaO.$$

Thus 1 mole of $CaCO_3$ produces 1 mole of CO_2. Moles of $CaCO_3$ = Moles of CO_2 produced $= 5/22.4 = 0.223$ moles. Molecular weight of $CaCO_3 = 40 + 12 + 48 = 100$. Weight of $CaCO_3$ required $= 0.223 \times 100 = 22.3$ grams.

8. The equation for the reaction is:

$$2\ C_2H_6 + 7\ O_2 = 4\ CO_2 + 6\ H_2O$$

Since volumes are in same ratio as coefficients,

$$\frac{2}{7} = \frac{15}{x}$$

$$x = 52.5 \text{ liters of oxygen.}$$

9. In PH_3, 31 parts by weight of P are combined with 3 parts by weight of H. Therefore equivalent weight of phosphorus is:

$$\frac{31}{3} = \frac{x}{1}$$

$$x = 10.3.$$

10. Step 1:

$$\frac{89.8}{10.2} = \frac{x}{8}$$

$$x = 70.43 \text{ (equivalent weight of element)}$$

Step 2:

$$\frac{6.2}{0.0305} = 203 \quad \text{(approximate atomic weight of element)}$$

Step 3:

$$\frac{203}{70.43} = 2.88 \text{ (approximate valence number)}$$

Nearest whole number is 3 (correct valence number)

Step 4: Exact atomic weight of the element is:

$$70.43 \times 3 = 211.3.$$

Problem Set No. 8

1. Room temperature must be above the critical temperature. Therefore argon, carbon monoxide, helium, hydrogen, nitrogen, and oxygen are gases at room temperature.

2.

(a) $\dfrac{-78.5 + 273}{31.1 + 273} = 0.64$ (b) $\dfrac{-34.0 + 273}{144.0 + 273} = 0.57$

(c) $\dfrac{78.5 + 273}{243.1 + 273} = 0.68$ (d) $\dfrac{-252.8 + 273}{-239.9 + 273} = 0.61$

(e) $\dfrac{-183.0 + 273}{-118.8 + 273} = 0.58$ (f) $\dfrac{100.0 + 273}{374.0 + 273} = 0.58$

3. Despite the extensive variation in the properties of the substances considered in Problem 2, the ratio is in remarkable agreement for each substance. The absolute normal boiling point seems to be approximately two-thirds of the absolute critical temperature.

4. The temperature indicated on the thermostat is $180°$ F. The boiling point of water at the top of Pike's Peak is about $62°$ C., or on the Fahrenheit scale:

$$F = 9/5\ C + 32$$
$$= 9/5 \times 62 + 32 = 144° \text{ F.}$$

Thus the water would boil below the operating temperature of the thermostat. Therefore the thermostat should be removed to permit circulation of the water at this high elevation.

5. To melt the ice: $25 \times 79.7 = 1992.5$ cal.
   To heat water to $100°$: $100 \times 25 = 2500.0$ cal.
   To boil the water: $25 \times 540 = \underline{13500.0}$ cal.
          Total 17992.5 cal.

Problem Set No. 9

1. (a) $\dfrac{0.02}{0.080} = 0.25$ M.

(b) $\dfrac{0.234}{58.5 \times 0.050} = 0.08$ M.

(c) $\dfrac{222}{111 \times 4} = 0.5$ M.

2. (a) $\dfrac{2 \times 0.2}{0.200} = 2.0$ N.

(b) $\dfrac{0.2}{0.200} = 1.0$ N.

(c) Equivalent weight of K_2CO_3 is ½ the molecular weight, or 69.

Normality $= \dfrac{2.76}{69 \times 0.400} = 0.1$ N.

(d) Equivalent weight of $Al_2(SO_4)_3$ is ⅙ of molecular weight, or 57.

Normality $= \dfrac{6.84}{57 \times 0.250} = 0.48$ N.

3. (a) $\dfrac{4.0}{40 \times 0.400} = 0.25$ m.

(b) $\dfrac{333}{111 \times 6} = 0.5$ m.

4. (a) $\dfrac{6.0}{256} \times 100 = 2.3\%$ (Note that in dilute solutions of solids in liquids, the solid dissolves without appreciable change in the volume of the solution.)

(b) $\dfrac{6.0}{120 \times 0.250} = 0.2$ M.   (c) $0.2 \times 2 = 0.4$ N.

5. $\dfrac{10}{50} \times 100 = 20\%$.

6. $\dfrac{34.0}{170 \times 0.750} = 0.267$ M.

7. The equivalent weight of $AlCl_3$ is ⅓ the molecular weight, or 44.5.

Normality $= \dfrac{2.67}{44.5 \times 0.400} = 0.15$ N.

8. $C_iV_i = C_fV_f$

$0.8 \times 35 = 0.5 \times X$

$X = \dfrac{0.8 \times 35}{0.5} = 56$ ml.

$56 - 35 = 21$ ml. of water to be added.

9. $N_1V_1 = N_2V_2$

$24.0 \times 0.1 = X \times 25.0$

$X = \dfrac{24.0 \times 0.1}{25.0} = 0.096$ N.

10. $N_s \times V_s$ (in liters) $= \#$ of equiv.

$X \times 0.020 = \dfrac{0.285}{143.5}$

$X = \dfrac{0.285}{143.5 \times 0.020} = 0.099$ N.

Problem Set No. 10

1. From Equation 15:

$$m = \dfrac{(w) \times (M) \times (P)}{(W) \times (P - p)}$$

$$m = \dfrac{23 \times 18 \times 17.5}{500 \times 0.16} = 90.56.$$

2. From Equation 17:

$$B - b = \dfrac{0.52\, w}{m}$$

$$w = 12.4 \times \dfrac{1000}{200} = 62.$$

For $C_2H_6O_2$, $m = 24 + 6 + 32 = 62$.

So, $B - b = \dfrac{0.52 \times 62}{62} = 0.52.$

Therefore the solution boils at $100.52°$ C.

3. From Equation 18:

$$m = \dfrac{0.52\, w}{B - b}$$

$$w = 15.5 \times \dfrac{1000}{100} = 155.$$

$$B - b = 101.3 - 100.0 = 1.3.$$

So, $m = \dfrac{0.52 \times 155}{1.3} = 62.$

4. From Equation 20:

$$m = \dfrac{1.86\, w}{F - f}$$

$$w = 11.5 \times \dfrac{1000}{100} = 115 \text{ g}.$$

$$F - f = 0 - (-2.325) = 2.325°.$$

So, $m = \dfrac{1.86 \times 115}{2.325} = 92.$

5. The ratio of the weight of glycol to the weight of water is:

$$\dfrac{1.26}{1} \times \dfrac{6}{12} = \dfrac{1.26}{2}.$$

Therefore, $w = 1.26 \times \dfrac{1000}{2} = 630$ g.

From Equation 20:

$$F - f = \frac{1.86\ w}{m}$$

$$F - f = \frac{1.86 \times 630}{62} = 18.9°\ C.$$

The solution freezes at $-18.9°$ C. We then convert this to °F.:

F = 9/5 C + 32.
$$F = 1.8 \times (-18.9) + 32 = -2°\ F.$$

Problem Set No. 11

1. Each mole of $CaCl_2$ produces 3 moles of ions. Therefore,

$$F - f = 3 \times 1.86 \times m$$

$$= 3 \times 1.86 \times \frac{222}{111} = 11.16°.$$

Therefore the solution freezes at $-11.16°$ C.

2. (a) $MgCl_2 = Mg^{++} + 2\ Cl^-$
   $(Mg^{++}) = 0.05$ M
   $(Cl^-) = 0.10$ M

   (b) $H_2SO_4 = 2\ H^+ + SO_4^{--}$
   $(H^+) = 0.50$ M
   $(SO_4^{--}) = 0.25$ M

   (c) $Fe_2(SO_4)_3 = 2\ Fe^{+++} + 3\ SO_4^{--}$.
   $(Fe^{+++}) = 0.2$ M.
   $(SO_4^{--}) = 0.3$ M.

3. (a) $K_2CO_3 + CaCl_2 = \underline{CaCO_3} + 2\ KCl$.
   (Forms precipitate)

   (b) $2\ HNO_3 + Na_2SO_3 = H_2SO_3 + 2\ NaNO_3$.
   (Forms weak electrolyte)

   (c) $H_2SO_4 + 2\ AgNO_3 = \underline{Ag_2SO_4} + 2\ HNO_3$.
   (Forms precipitate)

   (d) $HNO_3 + Na_2SO_4 =$ No reaction.

4. $$HNO_2 = H^+ + NO_2^-$$

   $$\frac{(H^+) \times (NO_2^-)}{(HNO_2)} = K$$

   $$\frac{(0.0063)^2}{0.1} = 4.0 \times 10^{-4}.$$

5. $$\frac{X^2}{0.1} = 1.8 \times 10^{-5}$$

   $$X^2 = 1.8 \times 10^{-6}$$

   $$X = 1.3 \times 10^{-3}\ M.$$

6. $$\frac{1.3 \times 10^{-3} \times 10^2}{10^{-1}} = 1.3\%.$$

7. $$\frac{(H^+) \times (CN^-)}{(HCN)} = 4 \times 10^{-10}$$

   $$\frac{X^2}{0.1} = 4 \times 10^{-10}$$

   $$X^2 = 4 \times 10^{-11} = 40 \times 10^{-12}$$

   $$(H^+) = X = 6.3 \times 10^{-6}$$

   $$pH = 6 - \log 6.3 = 6 - 0.80 = 5.2.$$

8. $$\frac{(NH_4^+) \times (OH^-)}{(NH_4OH)} = 1.8 \times 10^{-5}$$

   $$\frac{X^2}{0.1} = 1.8 \times 10^{-5}$$

   $$X^2 = 1.8 \times 10^{-6}$$

   $$(OH^-) = X = 1.3 \times 10^{-3}$$

   $$pOH = 3 - \log 1.3 = 3 - 0.11 = 2.89$$

   $$pH = 14 - 2.89 = 11.11.$$

9. $$\frac{(H^+) \times (C_2H_3O_2^-)}{(HC_2H_3O_2)} = 1.8 \times 10^{-5}$$

   $$\frac{X \times 0.01}{0.05} = 1.8 \times 10^{-5}$$

   $$(H^+) = X = 9.0 \times 10^{-5}$$

   $$pH = 5 - \log 9.0 = 5 - 0.95 = 4.05.$$

10. (a) $KNO_3 = K^+ + NO_3^-$
    $H_2O = OH^- + H^+$
    All strong electrolytes; no hydrolysis. Solution remains neutral.

    (b) $KNO_2 = K^+ + NO_2^-$
    $H_2O = OH^- + H^+$
    Forms HNO_2; therefore solution becomes basic.

    (c) $NH_4NO_3 = NH_4^+ + NO_3^-$
    $H_2O = OH^- + H^+$
    Forms NH_4OH; solution becomes acidic.

    (d) $Na_2SO_3 = 2\ Na^+ + SO_3^{--}$
    $2\ H_2O = 2\ OH^- + 2\ H^+$
    Forms H_2SO_3; solution becomes basic.

    (e) $FeCl_3 = Fe^{+++} + 3\ Cl^-$
    $3\ H_2O = 3\ OH^- + 3\ H^+$
    Precipitates $Fe(OH)_3$; solution becomes acidic.

11. Solubility $= 0.009$ g./liter $= \dfrac{0.009}{58}$ moles/liter $=$

1.6×10^{-4} M.

$$Mg(OH)_2 = Mg^{++} + 2\,OH^-$$
$$(Mg^{++}) \times (OH^-)^2 = K_{sp}$$
$$(1.6 \times 10^{-4}) \times (3.2 \times 10^{-4})^2 = K_{sp} =$$
$$1.6 \times 10^{-11}.$$

12.
$$AgC_2H_3O_2 = Ag^+ + C_2H_3O_2^-$$
$$(Ag^+) \times (C_2H_3O_2^-) = 2 \times 10^{-3}$$
$$X^2 = 2 \times 10^{-3} = 20 \times 10^{-4}$$
$$(Ag^+) = (C_2H_3O_2^-) = X = 4.4 \times 10^{-2}\ M.$$

13.
$$MgCO_3 = Mg^{++} + CO_3^{--}$$
$$(Mg^{++}) \times (CO_3^{--}) = 1 \times 10^{-5}$$
$$X \times 0.05 = 1 \times 10^{-5}$$
$$(Mg^{++}) = X = 2 \times 10^{-4}\ M.$$

14. (a) $MgCO_3 = Mg^{++} + CO_3^{--} \ldots + 2\,H^+ = H_2CO_3$. So salt dissolves.

(b) $AgCl = Ag^+ + Cl^- \ldots + H^+ =$ nothing. Salt does not dissolve.

(c) $Cu(OH)_2 = Cu^{++} + 2\,OH^- \ldots + 2\,H^+ = 2\,H_2O$. So salt dissolves.

(d) $BaSO_4 = Ba^{++} + SO_4^{--} \ldots + 2\,H^+ =$ nothing. Salt does not dissolve.

15.
$$Ca(OH)_2 = Ca^{++} + 2\,OH^-$$
$$(Ca^{++}) \times (OH^-)^2 = 8 \times 10^{-6}.$$

$$\frac{X}{2} \times (X)^2 = 8 \times 10^{-6}$$

$$\frac{X^3}{2} = 8 \times 10^{-6}$$

$$X^3 = 16 \times 10^{-6}$$
$$(OH^-) = X = 2.5 \times 10^{-2}\ M$$
$$pOH = 2 - \log 2.5 = 1.60$$
$$pH = 14 - 1.60 = 12.40$$

Problem Set No. 12

1. (a) Oxidation-reduction. (b) No valence change. (c) Oxidation-reduction.

2. (a) $H_2S + I_2 = S + 2\,I^- + 2\,H^+$. (b) Oxidizing agent is I_2. (c) I_2 is reduced.

3. (a) $2\,Cu^{++} + 2\,I^- = I_2 + 2\,Cu^+$. (b) Reducing agent is I^-. (c) I^- is oxidized. (d) This would not balance net charge.

4. (a) $2\,MnO_4^- + 5\,Sn^{++} + 16\,H^+ = 2\,Mn^{++} + 5\,Sn^{++++} + 8\,H_2O$. (b) The oxidizing agent is MnO_4^-. (c) Sn^{++} is oxidized.

5. (a) $Cr_2O_7^{--} + 6\,Fe^{++} + 14\,H^+ = 2\,Cr^{+++} + 6\,Fe^{+++} + 7\,H_2O$. (b) The reducing agent is Fe^{++}. (c) $Cr_2O_7^{--}$ is reduced.

Problem Set No. 13

1. (a) The more positive half-reaction proceeds in each case.

At anode: $2\,H_2O = O_2 + 4\,H^+ + 4\,e^-$

At cathode: $2\,H_2O + 2\,e^- = H_2 + 2\,OH^-$

(b) Net reaction is sum of half-reactions with electrons balanced. Doubling the coefficients in the second reaction and adding both reactions in part (a), we get: $2\,H_2O = 2\,H_2 + O_2$.

2. Noting that the cathode reaction has already been turned around, the minimum voltage will be the sum of the two emfs.

(a) $1.358 + 2.711 = 4.07$ volts.

(b) Since a dry cell produces only 1.5 volts, it will not do.

(c) Since an automobile battery produces 12 volts, it will do.

3. Wt. $= \dfrac{20 \times 30 \times 3600}{96500} = 22.38$ g.

4. Time $= \dfrac{96500 \times 40}{23 \times 20} = 8{,}391$ sec.

5. Volume $= \dfrac{35.5 \times 15 \times 3600 \times 22.4}{96500 \times 71} = 6.267$ liters.

6. (a) $Mg + NiCl_2 = MgCl_2 + \underline{Ni}$

(b) $3\,H_2 + 2\,AuCl_3 = 6\,HCl + 2\,\underline{Au}.$

(c) $Cu + ZnCl_2 =$ No reaction.

(d) $Ag + HCl =$ No reaction.

(e) $2\,Al + 3\,CuSO_4 = Al_2(SO_4)_3 + 3\,\underline{Cu}.$

(f) $Cu + 2\,AgNO_3 = Cu(NO_3)_2 + 2\,\underline{Ag}.$

7. (a) Half-reactions: $Al = Al^{+++} + 3\,e^-$
$$(+1.66\ v.)$$
$$Pb = Pb^{++} + 2\,e^-$$
$$(+0.126\ v.)$$

(b) Total emf: $1.66 - 0.126 = 1.534$ volts.

(c) Oxidizing half-reaction absorbs electrons. Therefore:
$$Pb = Pb^{++} + 2\,e^-.$$

8. (a) Half-reactions: $Pb = Pb^{++} + 2\,e^-$
$$(+0.126\ v.)$$
$$Ag = Ag^+ + e^-$$
$$(-0.7996\ v.)$$

(b) Total emf: $0.1263 - (-0.7996) = 0.9259$ volts.

(c) Reducing half-reaction gives off electrons. Therefore:
$$Pb = Pb^{++} + 2\,e^-.$$

(Contrast this with the results in Problem 7)

9. (a) At the anode, electrons are given off. Therefore, iron is the anode in the Edison cell.

(b) Total emf: $0.877 - (-0.49) = 1.367$ volts.

10. (a) Half reactions:

$$Fe^{++} = Fe^{+++} + e^- \quad (-0.770 \text{ v.})$$
$$2\,Cr^{+++} + 7\,H_2O = Cr_2O_7^{--} + 14\,H^+ + 6\,e^-$$
$$(-1.33 \text{ v.})$$

(b) Total emf: $-0.770 - (-1.33) = 0.560$ volts.

Problem Set No. 14

1. In each case divide the density by 1.293. N_2: 0.9675. O_2: 1.1052. CO_2: 1.5290. H_2O: 0.6218. H_2: 0.0696. O_3: 1.6582. Air: 1.0000.

2. From Table XIV, v. p. at $20° = 17.5$ torr; v. p. at $10° = 9.2$ torr

$$\frac{9.2}{17.5} \times 100 = 52.6\%.$$

3. From Table XIV: v. p. at $29° = 30.0$ torr $30.0 \times 89\% = 26.7$ torr. From Table XIV, 26.7 torr gives a dew point of 27° C. or 80.6° F.

4. On page 66.

5. $+7$.

6. Nitrogen, oxygen, hydrogen, helium, argon, neon, krypton, xenon, and possibly a bit of radon.

7. He, because molecular weight equals atomic weight.

8. A person, and the parts of his body, are cooled by moving through air because of increased rate of evaporation. Fast-moving bodies are actually heated, by friction with the air. Meteorites, on striking the atmosphere, are heated until they burn brilliantly with oxygen.

9. The dissolved carbon dioxide is liberated more readily by the shaking.

10. CO_2 comes off as a result of the action of citric acid on sodium bicarbonate.

Problem Set No. 15

1. Relatively inert—less active than iodine.

2. $\qquad 2\,KHF_2 = 2\,K + H_2 + 2\,F_2.$

3. In solution, HF is a weak electrolyte and therefore supplies fewer hydrogen ions per unit concentration than the strong electrolyte HCl. The

specific property of HF to attack glass has nothing to do with its acid characteristics.

4. At anode: Br_2. At cathode: H_2. Resulting solution: NaOH.

5. Probably the HCl, which forms in light or under the heat of the iron, attacks the fibers to produce the stains.

Problem Set No. 16

1. (a) ZnTe. (b) H_2Se. (c) H_2SeO_4. (d) $Na_2S_2O_3$.

2. $\qquad Zn + Te = ZnTe.$
$$ZnTe + 2\,HCl = ZnCl_2 + H_2Te.$$

3. Two or more forms of the same element having different properties as a result of different molecular or crystalline structures.

4. (a), (c), and (e) are weak. (b), (d), and (f) are strong.

5. (a) S is oxidized and O_2 is reduced.

(b) S in SO_3^{--} is oxidized and O_2 is reduced.

(c) S in SO_3^{--} is oxidized and free S is reduced. (Note: in this case the free S behaves just like O_2 in the previous equation and has a valence number of -2 in the thiosulfate ion. The S in the sulfite ion is oxidized from a $+4$ to a $+6$ valence number in the thiosulfate ion.)

Problem Set No. 17

1. Fish is a source of phosphate which is required in brain and nerve tissue.

2. $\qquad 2\,Sb_2S_3 + 9\,O_2 = 2\,Sb_2O_3 + 6\,SO_2$
$$Sb_2O_3 + 3\,C = 3\,CO + 2\,Sb.$$

3. $\qquad 4\,As + 5\,O_2 = 2\,As_2O_5$
$$As_2O_5 + 3\,H_2O = 2\,H_3AsO_4.$$

4. (a) H_3AsO_3. (b) H_3SbO_3. (c) H_3SbO_4.

5. Bicarbonate ions from the sodium bicarbonate react with hydrogen ions from the monosodium phosphate to produce carbon dioxide, which forms throughout the batter and causes it to rise. The equation is:

$$H^+ + HCO_3^- = H_2O + CO_2.$$

Problem Set No. 18

1. (a) Carbon monoxide. (b) Carbon dioxide.

2. Both producer gas and water gas are made from

steam and coke, but air is also used in making producer gas. This introduces large amounts of incombustible nitrogen, which lowers the heating value of the producer gas per unit volume.

3. Elements present in window glass are: oxygen, silicon, sodium, magnesium, calcium, aluminum, and iron.

4. Peach kernels contain deadly hydrogen cyanide. If a sufficient number of them are eaten, violent illness or even death can result.

5. Dry plant matter contains 44% carbon. Carbon dioxide contains only 27% carbon. Thus a substance less rich in carbon is being removed from the decaying plant matter.

Problem Set No. 19

1. $$2 KOH = H_2 + O_2 + 2 K.$$

2. The alkali metals react chemically by losing the outermost electron. In the heavier members of this family, this negative electron is farther removed from the positive nucleus and thus is given up more readily.

3. The equation is:

$$\overset{4.6}{\underset{46}{2 Na}} + 2 H_2O = 2 NaOH + \overset{X}{\underset{2}{H_2}}$$

Therefore:

$$\frac{4.6}{46} = \frac{X}{2}$$

$$X = \frac{4.6 \times 2}{46} = 0.2 \text{ g. of hydrogen.}$$

4. Ca^{++} from *limestone* goes into the formation of $CaCl_2$. Cl^- from *salt* goes into the formation of $CaCl_2$. *Water* is not recovered in the process.

5. Ammonium hydroxide is a solution of the gas ammonia in water. The solubility of all gases decreases as the temperature is increased. Therefore heating ammonium hydroxide will drive off ammonia gas from the solution.

Problem Set No. 20

1. (a) Heating carnallite to its melting point will drive off the water of crystallization as steam, so no hydrogen will be present in the fused bath to be a by-product.

(b) KCl: $2.924 - (-1.358) = 4.282$ volts.

$MgCl_2$: $2.37 - (-1.358) = 3.728$ volts.

(c) If the voltage during electrolysis is held below 4.28 volts, potassium cannot deposit.

2. $$Al_2(SO_4)_3 = 2 Al^{+++} + 3 SO_4^{--}$$
$$6 H_2O = 6 OH^- + 6 H^+$$
$$\overset{\|}{2 \underline{Al(OH)_3}}$$

3. (a) $NaAl(SO_4)_2 \cdot 12 H_2O$.

(b) Alkaline earth metals are divalent. Alums contain only monovalent and trivalent ions. Thus no alkaline earth metals are found in alums.

4. Be: $3 \times 9 = 27$
   Al: $2 \times 27 = 54$
   Si: $6 \times 28 = 168$ % Be $= \frac{27}{537} \times 100 = 5\%$.
   O: $18 \times 16 = \underline{288}$
   537

5. The chemical activity of metals depends upon their ability to lose electrons. As is the case with the alkali metals, the heavier alkaline earth metals lose electrons more easily because the electrons are farther from the nucleus of the atoms. Therefore radium, barium, and strontium will be *above* calcium in the activity series in this order: radium, barium, strontium, calcium, magnesium.

Problem Set No. 21

1. Hematite—70%; Magnetite—72%; Pyrites—47%; Siderite—48%.

2. See page 95.

3. (a) −3; (b) −1; (c) −2.

4. See page 144.

5. (a) $Cr_2O_3 + 2 Al = Al_2O_3 + 2 Cr$.

(b) $3 V_2O_5 + 10 Al = 5 Al_2O_3 + 6 V$.

Problem Set No. 22

1. Malachite, $Cu_2(OH)_2CO_3$. The copper is dissolved from piping by the water containing dissolved carbon dioxide.

2. By using the impure copper as the anode, copper and all metals above it in the activity series present in the anode go into solution as ions. All metals below copper in the activity series simply fall to the bottom of the bath as the anode decomposes. At the cathode, only copper, which is the least

active metal in solution, can deposit. Thus copper is effectively separated from all other metals.

3. $2 ZnS + 3 O_2 = 2 ZnO + 2 SO_2.$

$$ZnO + C = CO + Zn.$$

4. (a) Copper—wiring and electrical equipment.

   (b) Tin—Babbitt and bearing metals.

   (c) Mercury—mercury fulminate.

5. Because it is a source of poisonous lead salts formed when water containing dissolved oxygen dissolves lead.

Problem Set No. 23

1. A sacrificial anode must be chemically more active than the metal it protects. Since the noble metals are very inert, they would not serve as sacrificial anodes.

2. See page 147.

3. See pages 156, 157 for the two definitions.

4. Silver is removed from photographic film and printing paper as sodium silver thiosulfate in the fixing process. Since silver is much less active than sodium, silver would easily be obtainable as a cathode deposit by subjecting this solution to electrolysis.

5. Lanthanides: Lanthanum, Cerium, Praseodymium, Neodymium, Promethium, Samarium, Europium, Gadolinium, Terbium, Dysprosium, Holmium, Erbium, Thulium, Ytterbium, Lutetium.
   Actinides: Actinium, Thorium, Protactinium, Uranium, Neptunium, Plutonium, Americium, Curium, Berkelium, Californium, Einsteinium, Fermium, Mendelevium, Nobelium, Lawrencium.

Problem Set No. 24

1. ⟨◯⟩—OH

2. $C_2H_5OH + CH_3\overset{O}{\overset{\|}{C}}OH \rightleftharpoons CH_3\overset{O}{\overset{\|}{C}}OC_2H_5 + H_2O$

3. $\dfrac{1}{22.4} = \dfrac{1.88}{X}$

   $X = 42.1$ g. in 1 mole
   $0.856 \times 42.1 = 36.0$ g. carbon

$0.144 \times 42.1 = 6.1$ g. hydrogen
$36.0/12 = 3$ g.-atoms carbon
$6.1/1.01 = 6$ g.-atoms hydrogen
Therefore the molecular formula is C_3H_6.

4. $1 - d, \quad 2 - e, \quad 3 - c, \quad 4 - a, \quad 5 - b.$

5.
$$\begin{array}{ccccc} & H & H & H & \\ & | & | & | & \\ H- & C- & C- & C- & H. \\ & | & | & | & \\ & H & Cl & H & \end{array}$$

Problem Set No. 25

1. The atomic weights on both sides of the equation must balance. Therefore X must be written ^1X. According to the periodic table, only hydrogen matches ^1X. Therefore X = hydrogen, H.

2. In fusion, nuclei collide and fuse. In fission, nuclei split apart or disintegrate. In an ordinary chemical reaction the nuclei stay intact and the outer electrons surrounding the nucleus are involved in chemical reaction.

3. (a) From the nucleus.

4. Neutron-induced reactions may result in both heavier and lighter elements.

5. The energy produced by the sun involves fusion or thermonuclear reactions.

Problem Set No. 26

1. (a) Hydroelectric.

   (c) Wood.

2. True. They contain carbon and hydrogen, which oxidize or burn to CO_2 and H_2O.

3. Solar energy strikes silicon photovoltaic cells which provide the electricity to electrolyze water.

4. Refer to Table XLVIII.

$$\frac{16+3}{79} \times 100 = 24.1\% \text{ in 1980.}$$

$$\frac{71+18}{143} \times 100 = 62.2\% \text{ in 2020.}$$

5. $\dfrac{34.2}{0.252} \times 2 = 271$ BTU/mole.

Problem Set No. 27

1. 0.05 ppm $= 0.05 \times 10^{-6} = 5 \times 10^{-8}$ grams lead per gram of water.

2. (b) automobiles.

3. The rain is acidic; it contains sulfuric acid from stack gases, much of which is from coal-fired electric plants.

4. It is stable. It is situated away from rivers, lakes, and (it is to be hoped) underground water streams, and it is immobile.

5. Yes. CaO is basic, like MgO, and neutralizes acidic oxides such as SO_2 and SO_3.

Problem Set No. 28

1. (a) Carbohydrate.
   (b) Fat.
   (c) Protein.
   (d) Fat.
   (e) Fat.

2. (a) Polymer.
   (b) Monomer.
   (c) Monomer.
   (d) Polymer.
   (e) Monomer.

3. $2 C_{17}H_{35}COONa + Ca^{++} = (C_{17}H_{35}COO)_2Ca + 2 Na^{++}$.
   $(C_{17}H_{35}COO)_2Ca$ is insoluble and forms a ring around the bathtub.

4. (c) Animals and plants.

5. $-NH_2$. Remember that NH_4OH is a base. It can be written as $NH_2^- \, H_3O^+$. Therefore, NH_2^- can be considered to be the basic portion of NH_4OH as well.

CONCISE GLOSSARY OF CHEMICAL TERMS

Absolute Temperature—Temperature scale whose degrees are the same as Celsius but whose zero is 273 degrees below that of Celsius.

Absolute Zero—Zero on the absolute or Kelvin scale; the lowest temperature theoretically possible.

Acid—A compound which yields hydrogen ions in solution.

Acid Rain—Rain with a pH of less than 7, due to its absorption of acid oxides such as sulfur dioxide, carbon dioxide, and nitric oxide.

Activity—Ease of undergoing chemical change.

Allotropic Forms—Forms of an element having different molecular structure, causing differences in chemical and physical properties.

Alloy—Metallic solid produced by dissolving two or more molten metals in each other, and whose properties differ from those of its constituents.

Amalgam—An alloy of a metal and mercury.

Amphoteric—Acting either as acid or base.

Analysis—The determination of the composition of a substance; the decomposition of a substance.

Anion—Ion possessing a negative electrical charge; ion which will migrate to the anode.

Anode—The positive plate in an electrolytic system; the negative plate in a voltaic system; the plate where oxidation takes place and where electrons are lost.

Aqua Regia—A highly corrosive solution consisting of 3 parts hydrochloric acid and 1 part concentrated nitric acid.

Asteroid—One of many small rocky planets with orbits between Mars and Jupiter.

Atom—The smallest quantity of an element that enters into chemical combination. It has a very small, dense nucleus, positively charged. Negative particles called electrons revolve around the nucleus.

Atomic Number—The number of protons in the nucleus of an atom; the number of electrons surrounding the nucleus of an atom.

Atomic Weight—The mass of an atom compared with the mass of an atom of carbon taken as 12.

Base—The hydroxide of a metal. An alkali.

Biochemical or **Biological Oxygen Demand**—A method to estimate the amount of organic contamination of water, expressed as mg./l. of dissolved oxygen required by aerobic bacteria to decompose the organic matter.

Biochemistry—The chemistry of life and its constituents; the chemistry of plants and animals.

Boiling Point—The temperature at which the pressure of the vapor of a substance equals that of its surrounding atmosphere.

Bond—That which holds elements together in a compound; it may be electrovalent or covalent.

BTU—British thermal unit. One BTU equals the quantity of heat required to raise the temperature of one pound of water one degree Fahrenheit.

Carat—Unit of 24ths, by weight, used in expressing the composition of gold alloys; a unit (⅕ gram) for weighing precious stones.

Catalyst—A substance that changes the speed of a reaction without being itself permanently changed.

Cathode—The negative plate in an electrolytic system; the positive plate in a voltaic system; the plate at which electrons are taken up, and where reduction takes place.

Cation—Ion possessing a positive charge; ion which will migrate to the cathode.

Celsius—Thermometer scale on which the freezing point of water is zero and the boiling point is 100°.

Chemical Change—A drastic change in properties to the extent that a new substance is formed.

Chemical Properties—The characteristics inherent in a substance that give it the ability to change into a new and completely different substance.

Chemistry—The science that deals with the composition, behavior, and transformation of matter.

Chromosome—The heredity-bearing strings of genes and other protein components in the living cell. The human has 46 chromosomes, the horse, 60.

Codon—Three nucleotide pairs in a DNA strand. A few hundred to several thousand codons constitute a gene.

Combustion—Burning; oxidation with the emission of heat and light (flame).

Compound—A pure substance consisting of chemically combined elements.

Covalence—A type of bonding resulting from the sharing of pairs of electrons between atoms.

Critical Temperature of a gas—Temperature above which the gas can not be liquefied.

Crystal—A solid in which the particles (molecules, atoms, ions) are arranged in a definite pattern.

Decomposition—The process of breaking down a substance into simpler ones.

Deliquescence—The absorption of water from the air by a substance to the extent that the substance is dissolved in the absorbed water.

Density—The mass of a unit volume of a substance.

Desiccant—Drying agent; absorbs moisture from its surroundings.

Distillate—The condensed liquid obtained by a distilling process.

Electrode—The plate or terminal of a chemical electrical system.

Electrolyte—A substance which will conduct a current when melted or in solution. A substance decomposable by an electric current.

Electrolysis—Decomposition of a substance, melted or in solution, by passing an electric current through the liquid.

Electron—A constituent of an atom possessing a unit negative electrical charge and having negligible mass from a chemical point of view.

Electrovalence—A type of bonding in compounds in which oppositely charged ions are held together by attraction.

Element—The simplest form of matter.

Energy—The ability to do work. Heat and electricity are forms of energy. All chemical changes involve loss or gain of energy.

Environment—The atmosphere, bodies of water, and solid land that make up the earthly (and lunar) habitat of human beings, animals, and plants.

Equation—A concise statement of a chemical reaction using the symbols and formulas of the reactants and products.

Equilibrium—Condition in which two processes proceed simultaneously in opposite directions at the same rate.

Equivalent—A quantity of a substance equal in mass to its equivalent weight.

Equivalent Weight—The mass of a substance which has combined with or replaced 8 grams of oxygen.

Excess—An amount of a reactant beyond that theoretically required for a reaction.

Fahrenheit—Temperature scale on which the freezing point of water is 32° and the boiling point is 212°.

Filtrate—The liquid obtained after filtration.

Filtration—The process of separating a liquid from a solid by straining it through porous paper or similar material.

Flux—Material which permits another substance to melt more easily.

Formula—The composition of a substance indicated by symbols of each element present and subscript numbers showing the number of each type of atom involved.

Formula Weight—Sum of the atomic weights of all atoms in a formula; molecular weight.

Freezing Point—Temperature at which a liquid changes to a solid.

Fused—Molten.

Gene—A segment of DNA that can transmit a message that influences a chemical function finding expression in hereditary characteristics.

Hydrolysis—Reactions between ions of a salt and ions of water forming a solution which is either acidic or alkaline.

Hypothesis—An intelligent guess as to the nature of a phenomenon.

Inert—Having little tendency to undergo chemical change.

Ion—An atom which has acquired an electrical charge as a result of the gain or loss of electrons; a charged atom or radical.

Isotopes—Atoms of an element having the same chemical properties but differing in atomic weight.

Kilocalorie—One thousand calories of energy.

Landfill—A site on or in solid ground at a distance from any body of water above or below ground, for the disposal of wastes.

Law—A non-varying performance in nature.

Matter—Anything that occupies space and has mass.

Metal—Element which readily loses electrons to form positive ions; conductor of electricity.

Metalloid—Element possessing characteristics of both metals and nonmetals.

Mixture—A combination of substances held together by physical rather than chemical means.

Mole—A quantity of a substance equal in mass to its molecular (formula) weight.

Molecular Weight—The mass of a molecule compared to the mass of a carbon atom of 12 units mass; formula weight.

Molecule—Smallest particle of a compound capable of having the properties of the compound.

Neutralization—Removal of either acidity or alkalinity; reaction of an acid and a base to form a salt and water.

Neutron—A constituent of an atom possessing no electrical charge and having a mass of 1 compared to a carbon atom of 12 units mass.

Noble—Not chemically active; relatively inert.

Nonmetal—Element which readily gains electrons to form negative ions; nonconductor of electricity.

Nonrenewable Energy—Energy that can not be replenished, such as energy from petroleum and coal.

Nucleus—The kernel of an atom possessing essentially all of the mass of the atom and a positive electrical charge.

Oxidation—Process involving loss of electrons by a substance; gain in positive valence number; combination with oxygen.

pH—Logarithm of the reciprocal of the molar concentration of the hydrogen ion; a scale indicating the acidity of a solution; if pH is less than 7, solution is acidic; if pH is 7, solution is neutral; if pH is greater than 7, solution is alkaline.

Phenomenon—A particular natural event.

Photovoltaic Cell—A thin wafer of inorganic material, such as silicon, that converts sunlight impinging on it into electricity.

Physical Change—An alteration of the properties of a substance that does not affect the substance itself.

Physical Properties—Inherent characteristics that describe matter as it is.

Polar—Possessing poles or centers of both positive and negative electrical charge at two different locations in its structure.

Precipitate—A solid that forms in and settles out from solution.

Products—The substances formed as a result of a chemical change.

Proton—A constituent of an atom possessing a unit positive electrical charge and having a mass of 1 compared to a carbon atom of 12 units mass; a hydrogen ion.

Quad—10^{15} or 1 quadrillion BTU of energy.

Radical—A group of elements bonded together which behave chemically as a single atom.

Radioactivity—The emission of alpha, beta, and gamma rays from a substance through the breakdown of the nuclei of its atoms.

Reaction—A chemical change.

Reactants—The substances that react with one another in a chemical change.

Reduction—Process involving gain of electrons by a substance, decrease in valence number, or combination with hydrogen.

Renewable Energy—Energy that can be replenished, such as energy derived from dams, agricultural crops, and the wind.

Research—Directed investigation of natural phenomena to acquire deeper understanding of the behavior of nature or to develop new products.

Roasting—Heating a compound in air to convert it to an oxide.

Salt—Compound which ionizes, but which produces neither hydrogen nor hydroxide ions in solution.

Saturated—Containing the maximum amount possible under the conditions.

Science—The study and the knowledge and understanding gained from the study of the behavior of nature.

Solubility—The extent to which a solute can dissolve in a solvent; the concentration of a saturated solution of a given solute.

Solute—That which is dissolved in a solvent.

Solution—Homogeneous, nonsettling mixture of two ingredients, solute and solvent.

Solvent—The medium in which a substance is dissolved.

Specific Gravity—The ratio of the density of a liquid or solid to that of water.

Stable—Relatively inert; hard to decompose.

Standard Conditions—0°C. and 760 torr pressure or their equivalent.

Standard Solution—Solution of accurately known concentration.

Strong Electrolyte—Substance that ionizes completely in water solution.

Substance—A definite variety of matter; it may be an element or a compound.

Symbol—A letter or letters representing the name of an element, one atom of it, and a quantity of it equal in mass to its atomic weight.

Synthesis—The formation of a compound by combining elements or more simple compounds.

Theory—A general explanation of related phenomena supported by evidence.

Toxic Chemicals—Poisonous chemicals which can cause injury through exposure to or ingestion of even small amounts.

Trace Metals—Metals essential to plant and animal nutrition in trace amounts, 1000 ppm. or less.

Valence—The tendency of elements to form compounds; a number indicating a charge on an ion or the number of pairs of electrons shared by one element with another; the bonding in a compound.

Vapor—A substance in the gaseous state, usually a gas in equilibrium with its liquid or solid, such as water vapor.

Volatile—Having a tendency to evaporate readily to form a vapor.

Water Gas—A mixture of carbon monoxide and hydrogen; it is an important fuel.

Weak Electrolyte—Substance that ionizes to only a slight extent in water solution.

INDEX